**SLOOPS
1926-1946**

CYGNET turning in the Mersey with the Wirral shoreline behind her, prior to entering the Gladstone Dock, possibly straight from her builders' yard. Oddly, she wears the funnel insignia of 45 EG, although she is not known to have served with this Group, possibly indicative of an early intention whilst still at the builders.

SLOOPS

A history of the 71 sloops built in Britain and Australia for the British, Australian and Indian Navies

1926-1946

Arnold Hague

INTRODUCTION

This book is written as a natural successor to **THE HUNTS, THE TOWNS** and **AMAZON to IVANHOE**, following in the style set by John English in the first of the trio. In addition to describing the philosophy of the type and its development over a period of fourteen years, I have attempted to give a brief history of each vessel. In order to keep the extent of the text within reasonable limits, it has been necessary to exercise brevity in dealing with the more mundane pre and post-war periods, to accord as much space as possible to the great involvement of the ships in the 1939-45 war.

Official records held at the Public Record Office at Kew, in the National Maritime Museum archives at Woolwich, and by the Naval Historical Branch of the Ministry of Defence (Navy) have been used extensively in the preparation of this book. In particular, the attribution of wartime submarine sinkings has depended greatly upon the ongoing work of the Foreign Documents Section of the Naval Historical Branch; this is reflected in the number of changes from the details announced in 1947 and frequently repeated in the last forty five years.

ACKNOWLEDGEMENTS

The assistance of David Brown, Head of Naval Historical Branch, Bob Coppock, Alan Francis and Mike McAloon also of that Branch, in making available documents and facts; Paul Kemp and David Hodges of the Photographic Departments of (respectively) the Imperial War Museum and the National Maritime Museum in searching out photographs and information to explain discrepancies in information, to Ian Buxton for the use of his detailed knowledge on shipbreaking, Roy Hall of Australia for much search for illustrations of Australian ships, is gratefully acknowledged. Finally, and in particular, the considerable assistance afforded by Michael Crowdy in ensuring that the presentation of this text is up to standard; a tedious and time consuming task for which I am most grateful.

PHOTOGRAPHS

Unless otherwise stated, all photographs are from the author's collection and originated from the now defunct Admiralty Archive. Such of that Archive as has survived is now in the hands of the Imperial War Museum or, occasionally, the National Maritime Museum. In a few cases where these Museums have supplied pictures direct from their collections, that fact is shown beneath the illustration, and all enquiries should therefore be directed to those sources. Photographs credited "WSPL" are the property of the Society from its general archive; copies of these photographs will become available in the normal course of offers to the membership of the Society.

Cover photograph: H.M.S. STARLING

© **1993 A. Hague**

ISBN 0 905617 67 3

**Published in 1993 by the World Ship Society,
28 Natland Road, Kendal LA9 7LT, England.**

All rights reserved. No part of this book may be reproduced or transmitted in any form or by any means electronic or mechanical including photocopying, recording or any information storage and retrieval system, official, public or personal, without prior permission in writing from the Publisher.

Companion Volumes: THE HUNTS: John English
                         THE TOWNS: Arnold Hague
                         AMAZON TO IVANHOE: John English
                         CONVOYS TO RUSSIA: Bob Ruegg and Arnold Hague

Contents

Introduction	4
Acknowledgements	4
Photographs	4
Sloops — A chronology	6
Royal and Commonwealth Navy Sloop Construction Post 1920	7
Glossary	8
Sloops: The 20th Century Sloop in the Royal Navy	9
The Revival of the title of "sloop"	9
Post 1920 developments	10
Design and development of the replacement ship	10
The progressive development of the minesweeping sloop	11
Further thoughts on minesweeper and sloop development	13
1930s proposals on sloop designs	14
Subsequent sloop development up to 1944	15
The BLACK SWAN and Modified BLACK SWAN classes	17
Wartime service	18
Post war service	19
The end of service	19
Overseas service	20
Sloop conversions	20
Armament changes	21
Locally entered personnel	22
BRIDGEWATER class	23
HASTINGS class	26
RIN HASTINGS class	31
SHOREHAM class	32
Repeat SHOREHAM class	38
GRIMSBY class	42
RAN GRIMSBY class	55
RIN INDUS class	60
BITTERN class	61
EGRET class	67
RIN Modified BITTERN class	71
BLACK SWAN class	76
Modified BLACK SWAN class	82
RIN Modified BLACK SWAN class	114
List of abbreviated titles for ship builders and engine makers	117
Pennant Numbers	118
Sloop disposals	119
Sinkings of Submarines	122
Index	124

SLOOPS - A CHRONOLOGY

This listing of the sloops constructed in British yards is given in chronological order of the date of instruction to commence work, ie "ordered". It is odd that the last two orders should be from a preceding Programme year to ships ordered prior to them, however the Programme Year has been confirmed from their SHIP'S BOOK entries. The explanation is cancellation of the original orders and later placement of the contracts with different builders.

Regrettably, no information is available on the Australian built vessels to include them in the listing; nor has it proved possible to complete the occasional gaps in the list of sponsors and order dates.

Name	Programme	Ordered	Launched by
BRIDGEWATER	1927	19.9.27	no ceremony
SANDWICH	1927	19.9.27	no ceremony
FOLKESTONE	1928	26.2.29	no ceremony
SCARBOROUGH	1928	26.2.29	no ceremony
HASTINGS	1928	4.4.29	Hon Alice Brand
PENZANCE	1928	4.4.29	Mrs Backhouse
HINDUSTAN			
BIDEFORD	1929		Hon Alice Brand
SHOREHAM	1929	23.9.29	Countess of Darnley
FOWEY	1929	4.12.39	Mrs Treffery
ROCHESTER	1929	30.4.30	Mayor, Mrs G Jenner
DUNDEE	1930	16.3.31	Mrs Round-Turner
FALMOUTH	1930		Viscountess Astor
MILFORD	1930	24.4.31	Mrs Reinold
WESTON	1930		Lady Addison
GRIMSBY	1931	1.11.32	Mrs Forbes
LEITH	1931	1.11.32	Mrs Pipon
LOWESTOFT	1932	1.5.33	Mrs Henderson
WELLINGTON	1932	1.5.33	Lady Fullerton
DEPTFORD	1933		Mayor, Mrs Pearson
LONDONDERRY	1933	1.3.34	Mrs Noble
INDUS			Lady Walwyn
ABERDEEN	1934	1.3.35	Lady Provost Mrs E Watt
FLEETWOOD	1934	1.3.35	Lady Plunkett-Ernle-Erle-Drax
ENCHANTRESS	1933		
STORK	1934		Mrs Edgee
BITTERN	1935	26.5.36	Lady Jean MacDonald
EGRET	1936	5.3.37	Lady Mitchelson
AUCKLAND	1936	5.3.37	Mrs W Denny
PELICAN	1937	19.3.37	Mrs W Rogers
BLACK SWAN	1937	1.1.38	Mrs Hogarth
FLAMINGO	1937	1.1.38	Mrs Parden
ERNE	1939	21.6.39	Mrs Furness
IBIS	1939	21.6.39	Mrs McGovern
JUMNA	1939	8.9.39	Mrs Rangasthan
SUTLEJ	1939	8.9.39	Mrs Amery
WHIMBREL	1940	13.4.40	Mrs White
WILD GOOSE	1940	13.4.40	Mrs Yarrow
WOODPECKER	1940	13.4.40	Mrs Greig
WREN	1940	13.4.40	Dame Vera Laughton-Matthews
GODAVARI	1940	29.8.40	Mrs Lail
NARBADA	1940	29.8.40	Mrs Amery
CHANTICLEER	1940	9.1.41	Lady Denny
CRANE	1940	9.1.41	Mrs Roberts
CYGNET	1940	27.3.41	Mrs Simpson
KITE	1940	27.3.41	Mrs Cooper
LAPWING	1940	27.3.41	Mrs Evelegh
LARK	1940	27.3.41	Mrs Phillips
MAGPIE	1940	27.3.41	Mrs Turner
PEACOCK	1940	27.3.41	Mrs Donaldson
PHEASANT	1940	27.3.41	Mrs Mitchell
REDPOLE	1940	27.3.41	Miss Lodge
STARLING	1940	18.7.41	Miss Featherstone
WOODCOCK	1940	18.7.41	Lady Hay

CAUVERY	1940	10.9.41	Lady Runganadhan
KISTNA	1940	10.9.41	Mrs Godfrey
ACTAEON	1941	3.12.41	
AMETHYST	1941	19.12.41	Miss Grange
HART	1941	18.12.41	Mrs Pitcher
HIND	1941	11.2.42	Mrs Denny
MERMAID	1941	11.2.42	Miss Gordon
MODESTE	1941	3.3.42	Miss Swindell
NEREIDE	1941	3.3.42	Miss Swindell
OPOSSUM	1941	12.8.42	Mrs Trinder
ALACRITY	1941	12.8.42	Miss Stewart
SPARROW	1940	8.12.42	Mrs Russell
SNIPE	1940	8.12.42	

ROYAL AND COMMONWEALTH NAVY SLOOP CONSTRUCTION
POST 1920

BRIDGEWATER class	BRIDGEWATER, SANDWICH
HASTINGS class	FOLKESTONE, HASTINGS, PENZANCE, SCARBOROUGH, HINDUSTAN
SHOREHAM class Note 1	BIDEFORD, FOWEY, ROCHESTER, SHOREHAM
FALMOUTH class Note 2	DUNDEE, FALMOUTH, MILFORD, WESTON
GRIMSBY class	GRIMSBY, LEITH, LOWESTOFT, WELLINGTON
Repeat GRIMSBY class	ABERDEEN, DEPTFORD, FLEETWOOD, LONDONDERRY
RAN GRIMSBY class	PARRAMATTA, SWAN, WARREGO, YARRA
RIN TYPE	INDUS
BITTERN class Note 3	BITTERN, ENCHANTRESS, STORK
EGRET class Note 4	AUCKLAND, EGRET, PELICAN
RIN Modified BITTERN class	GODAVARI, JUMNA, NARBADA, SUTLEJ
BLACK SWAN class	BLACK SWAN, ERNE, FLAMINGO, IBIS
Modified BLACK SWAN class Note 5	ACTAEON, ALACRITY, AMETHYST, CHANTICLEER, CRANE, CYGNET, HART, HIND, KITE, LAPWING LARK, MAGPIE, MERMAID, MODESTE, NEREIDE, OPOSSUM, PEACOCK, PHEASANT, REDPOLE, SNIPE, SPARROW, STARLING, WHIMBREL, WILD GOOSE, WOODCOCK, WOODPECKER, WREN
RIN Modified BLACK SWAN class	CAUVERY KISTNA

Notes.
1. Two additional ships for 1930 Programme cancelled.

2. WESTON originally to be named WESTON-SUPER-MARE.

3. ENCHANTRESS originally designated as BITTERN.

4. AUCKLAND ordered as HERON, renamed 23.8.37 prior to launch.

5. NONSUCH, NYMPHE, PARTRIDGE, WATERHEN and WRYNECK cancelled.

GLOSSARY

AA	Anti-Aircraft
ASDIC	The 1939/47 equipment now known as SONAR
A/S	Anti-Submarine
BISCO	British Iron and Steel Corporation, the supervisory body controlling the scrapping of warships post 1945
DC(s)	Depth charge(s)
DCT(s)	Depth Charge Thrower(s)
Director	The Gunnery director position, in sloops containing the rangefinder and topped, in wartime, by the 285 radar aerials
EF	Escort Flotilla
EG	Escort Group
ES	Escort Squadron
EV	Escort Vessel
FF	Frigate Flotilla
FS	Frigate Squadron
HA	High Angle, ie anti-aircraft capable
Hedgehog	An anti-submarine weapon that projected 24 bombs ahead of the ship to fall in an elipse, exploding on contact
HF/DF	Usually referred to as "Huff Duff". High Frequency Direction Finding equipment capable of fixing the bearing of a wireless transmission with great speed and accuracy. Responsible for the location and sinking of many submarines, and regarded as at least as valuable as radar, if not more so, in 1941/45
kts	Knots = the number of nautical miles covered in one hour
LA	Low Angle, ie a limited elevation weapon for surface warfare
mg	Machine Gun
M/S	Minesweeper or Minesweeping
oa	overall length, from the extremities of the ship
pdr	Pounder, a reference to the weight of shell fired by the gun
pp	between perpendiculars. A measure of length between verticals struck from the juncture of stem and keel, and the rudder post, ie the length involved in docking blocks
SHP	Shaft Horse Power, a measure of the power of the engine(s) installed in the ship
Squid	An ahead throwing A/S weapon consisting of a three barrelled mortar projecting depth fused bombs well ahead of the ship whilst still in ASDIC contact. Bombs fell in a triangular pattern designed to bracket the target ahead, astern and to one side. Ships mounting two Squid had the mountings adjusted to produce a double triangle with two bombs ahead, two astern and one on each side of the target
tph	tons per hour = the tonnage of fuel consumed in one hour at the speed quoted
wl	waterline length

Radar sets

271	The first effective surface search radar set designed for naval use. The aerial was enclosed in a multi-sided timber and plastic structure resembling a lantern, and called such in common usage
284	A gunnery ranging set
285	A gunnery ranging set, aerials in the form of "fishbones" mounted on top of the director tower
286	The first ship-borne radar, used for surface search and of dubious efficiency. An aerial, of open rod form and resembling a "mattress", was initially fixed below the masthead, later a rotating pattern at the masthead
291	The highly effective successor to 286, the aerial was a small, double, St Andrews Cross at the masthead
NOTE	The numbers of radar sets were frequently suffixed by a letter, indicating successive improvements of the set, usually in strength of signal and therefore range and definition

SLOOPS

The 20th Century Sloop in the Royal Navy

The title "Sloop" is an old and honoured term from the days of the sailing Navy. In the Victorian era it became associated with a small, relatively long endurance, steam warship with, initially, sail as auxiliary propulsion, which was extensively employed on distant Stations to supplement the small cruisers operated there; the smaller version of the type enjoyed the even more evocative term of "gunboat". The second half of the 19th Century history of the Royal Navy contains innumerable examples of the employment of these vessels overseas where they provided a reasonably economic example of seapower, in the colonial era. Construction of the type, which had been the subject of steady development over a long period, ceased in 1903 with the concentration of British maritime strength in Home waters and the increasing preoccupation with the probable conflict with the rapidly developing Kaiserliche Marine across the North Sea.

The type did not reappear in British service until 1916, when a quite different design appeared for duties totally unconnected with the holder of the original title. It is coincidental that the post-1919 duties of the survivors of this class included similar duties to the pre-1903 ships.

The revival of the title of "sloop"

The title was revived in 1916 and applied to the vessels of the initial FLOWER class of 1916 and its subsequent variations. The concept of the initial design was that of a general duties vessel combining the ability to carry out wire minesweeping, either singly or as a pair, to escort merchant ships against the rising submarine menace, to undertake extended anti-submarine patrols and, finally, to act as liberty boats for the Grand Fleet. This last requirement imposed the design requirement that the vessel had to have the ability to carry 1,000 libertymen, standing, under cover. This unusual proviso ensured that the design had a considerable margin of stability in the normal state and produced a surprisingly large hull for the remaining requirements, a factor which stood the ships in good stead later in their lives. The hull so designed could, until quite recently, be seen under the White Ensign on the Thames as both the drillships of London Division RNR were of this type. CHRYSANTHEMUM, although no longer an H M Ship, still remains in her original berth and was serving at sea as late as 1936.

GENTIAN, a FLOWER class sloop equipped for minesweeping: note the gallows aft. These ships, which served world wide in the sloop role post 1919 and required replacement in the late 1920s, gave rise to the BRIDGEWATER class.

BELVOIR, a twin screw HUNT showing the typical 1918 M/S rig with trawler type gallows aft. It was to replace ships of this type that the initial sloop classes were designed.

Post-1920 developments

After the end of hostilities in 1919, the term "sloop" was usually linked with another, more descriptive, title e.g. the HUNT class minesweepers were described as "minesweeping sloops" and, as a result of their wartime activities, the FLOWER class ships acquired the prefix "convoy" to their title in official practice.

In 1924, the condition of the remaining ships of the HUNT class was causing concern due to their general age and hard usage since construction (the class had been widely used in the very extensive post-war mine clearance programme and also as useful low cost general purpose vessels); and the Admiralty began to contemplate a replacement design for the mid/late 1920s. As always in peace time, with Treasury restrictions looming, a dual purpose vessel was sought, a replacement for the Flower class on distant Stations also being needed. The new design, therefore, had to satisfy the requirements of what we would now term a Fleet Minesweeper for war service and be capable of acting as a replacement sloop for peace time service in distant waters. Design papers make it quite clear that the minesweeper capability took precedence over all other factors, and the initial design is always referred to as a "minesweeping sloop" in correspondence.

Design and development of the replacement ship

The capability of Fleet Minesweeping — a term incidentally which was used to describe the ability both to sweep AND to proceed to a distant Station to carry out that duty relatively independent of outside support NOT, as is sometimes assumed, to sweep ahead of the Fleet - required that the ship should be able to proceed to a war station possibly well in excess of 1,000 miles away. There was a draught limitation of 9ft maximum, and requirements for a sweeping speed of 12 knots and an endurance of 3,000 miles at maximum sea speed. A sweep speed of 12 knots implied a sea speed of 16 knots maximum. Within these parameters, the requirements of a peace time sloop overseas had to be accommodated. One other, implied, requirement existed which exercised a great influence on design: the design of a minesweeper automatically required that the galley and associated spaces, the mess decks, heads and all like facilities, together with ready use water and victualling stores and supplies, MUST be above the waterline, ie accessible whether or not the ship was in potentially hazardous waters.

By October 1924 the requirement was refined to "combine the qualities of the sloop and the twin screw minesweeper" on a maximum length of 300ft and a draught of 8ft 6in, 5,000 miles endurance, twin screws and 18 knots sea speed. Even at this early date, it was also insisted that all guns must be High Angle(HA), ie. anti-aircraft, that ASDIC should be fitted if at all possible; and that internal combustion (diesel) engines should be considered. A month later, on 3 November 1924, it was decided to build two ships of 1,300 tons, 280ft overall, 18 knots, armed with two single 4in HA. One ship was to have geared steam turbines and one to have diesel propulsion, the intention being to run comparative trials. Within the month, the diesel proposal had to be discarded as no suitable machinery was available from British sources, such then being a Government requirement. In February 1925, the Admiralty was forced to accept that one of the two 4in mountings must be Low Angle (LA), as these were available from stock at some 20% of the cost of a new HA mounting, so the design was formalised as 280ft between perpendiculars, 36ft beam, 10ft 6in draught at deep displacement (which was set at 1,600 tons), two shaft geared steam turbine machinery producing 2,000 SHP. A reversion to two single 4 inch HA mountings was also called for, plus one of the new 0.5in multiple machine gun mountings then under development.

When the orders came to be placed for the 2 ships forming the first sloop class under consideration, the final design dispensed with the quadruple 0.5in machine gun mounting (lack of money had delayed its development so that it was not ready in time to be fitted), the after 4in mounting became an LA one from store and was fitted in the position originally allocated to the multiple machine gun. The 2 ships were ordered, as BRIDGEWATER and SANDWICH, to serve in the Persian Gulf in the sloop role.

BRIDGEWATER in her initial form, lying in Plymouth Sound shortly after completion. Note the simplicity of the design, the M/S winch and davits aft, and the solid bulwarks of the sweep deck.

A great deal of design effort had been expended in producing these two ships of a new type and design, most of it in fact devoted to their peace time role as sloops despite the Staff Requirement emphasis on minesweeping although this was constantly referred to. For example, sweeping gear was fitted and carried at all times, the anti-submarine armament (depth charges only at this stage) was manufactured but not fitted, the chutes for the charges being carried onboard as components in store, with the throwers laid aside ready for fitting at the appropriate dockyard where the ship served. In fact, the ships rarely if ever streamed a sweep operationally, at quarterly intervals for training purposes was the norm, and their major role was to be anti-submarine in wartime though it was to be ten years before this fact was officially recognised.

The progressive development of the minesweeping sloop

The initial class of two ships having proved satisfactory in service, receiving in fact considerable praise from the Royal Navy for the versatility of their design on a small hull, it was decided to proceed with a steady development programme. From the Service standpoint, not the least of the virtues of the new ships was their low financial cost, a major factor in the never-ending battle with the Treasury in an era of stringent economy. This outlook led to several points being overlooked, in particular the very tight accommodation provision.

Overseas service on detached duty placed unusual strains on accommodation. In addition to the constraints of minesweeping requirements on domestic accommodation, service in the Persian Gulf, Africa and China (three highly likely destinations) where there was little or no shore support, also required a slightly larger complement (a medical officer and SBAs had to be carried), also a properly fitted sick bay had to be provided in the total absence of shore hospital facilities. Provision also had to be made for local, native, personnel which implied separate messdecks, galley and heads. All this placed strains on a small hull, as the medical inspection on completion of the first ships found. The Senior Medical Officer noted in his report that "the Victorian Common Lodging House legislation specified 400 cubic feet of space per person accommodated", and he went on to comment that the capacity of the principal European messdeck "which he understood was to accommodate X ratings" meant that they would each have slightly in excess of 100 cubic feet of space; he made no other comment, presumably as such would have been superfluous in the circumstances.

SCARBOROUGH in her as built state 3.11.30.

Whatever the problems, the Admiralty obtained sanction to construct a further 4 ships, which became known as the HASTINGS class, while the Indian Government chose to use the basic design for the first post-war new building for the Royal Indian Marine, the ship completing as the HINDUSTAN. Alterations to the design of the new ships was minimal, being mainly confined to internal arrangements, exchange of the use of internal spaces etc, all aimed at improving habitability on overseas service and enhancing the minesweeping role, still officially described as "the primary function of all sloops in wartime". The HINDUSTAN differed in her internal arrangements from the four RN ships as her accommodation was designed to accommodate Indian ratings rather than British.

The HASTINGS class was followed by a further class of 4 ships (2 additional ships to have been ordered under the 1930 Estimates were cancelled) named as the SHOREHAM class, repeated by a further 4 ships

A splendid PR view of FOWEY visiting Fowey soon after completion. If nothing else it shows the complex rigging and W/T aerials of the early sloops.

A workmanlike MILFORD not long after completion, emphasising how basic the pre-war designs really were compared with those of the late 1930s.

WSPL

officially titled Repeat SHOREHAM although sometimes referred to, even in Admiralty documents, as the FALMOUTH class. Thus 14 modern units (excluding the Indian ship) were added to the Fleet, with both a meaningful peacetime role and certainly fully capable of their wartime duty, no mean effort at a time when disarmament and financial economy were international and national slogans.

Further thoughts on minesweeper and sloop development

By 1932 the Naval Staff were beginning to accept that the minesweeping and anti-submarine roles were becoming increasingly incompatible even at that early stage in the development of A/S warfare, and that the inevitable growth of the design and costs militated against its use as a minesweeper. Consequently a second design began development which would be smaller (and cheaper) than the current sloops, capable of mass production solely as a minesweeper. This design eventually became the HALCYON class minesweepers though even these ships were regarded by the Service as unnecessarily large and expensive for the minesweeper role. This in turn led to the design of the BANGOR class. It is worth noting that in the event the wheel turned full circle; the BANGOR design proved too small for the wartime sweeping role, especially when it embraced the unforeseen magnetic and acoustic requirements which had been quite unknown in the early 1930s and the eventual fleet minesweeper type was the magnificent ALGERINE which more nearly approached the HALCYON in its configuration. The HALCYONs, in their wartime service, spent almost as much time on anti-submarine work as they did on minesweeping and earned a high reputation in both roles.

Reverting to the principal sloop type, the acceptance that it was now to be considered "suitable as an A/S vessel for mass production in wartime" was too late to affect the design of the GRIMSBY class which followed on, except that the principle of upgrading armament led to the introduction of the 4.7in gun in lieu of the standard 4in used up till then. It was very quickly seen that this was a retrograde step when it became apparent that no HA mounting for that calibre would be available for the new ships. After a good deal of heart searching, the inevitable compromise had to be accepted, the 4.7in was fitted as an LA weapon in two positions, A & X, with a 3in HA fitted in B position. This was an unsatisfactory solution, requiring as it did stowage for two different calibres of shell, but was accepted pending the expansion of the basic design to accommodate the new twin 4in HA/LA mounting then under development. There was a divergence in this class from the usual pattern of British seaport names. One ship was destined to serve on the New Zealand Station, where she would have a strong contingent of local men onboard, hence the choice of the name WELLINGTON as being most appropriate. The 4 ships of the GRIMSBY class were followed by a repeat order of 4, entitled Repeat GRIMSBY, and by the adoption of the basic design by the Royal Australian Navy who also built four ships; these however, being spread over several years in building, emerged as four separate and distinct units. This last batch of 4 RN sloops therefore became the final development of the original design, future vessels were to follow a different design path as the

FLEETWOOD pre-war, displays the then new twin 4in HA armament on a sloop hull, with B position occupied by the derisory quadruple 0.5in mg. Minesweeping gear is still very much in evidence in this 4.37 photograph.

influence of the minesweeping ancestry declined and anti-submarine and anti-aircraft capability became prime requirements. A final ship built for the Royal Indian Navy, INDUS, was basically a repeat of the original HINDUSTAN as a commonality of spares and ships of similar internal layout was preferred in view of their local crews.

1930s proposals on sloop design

It is interesting to consider here a 1932 Staff concept for a new sloop, which somewhat counters the usually accepted view that the Royal Navy's Flag Officers were hidebound and thought only of the previous war. Two Flag Officers afloat, one Battle Squadron and one Cruiser Squadron commander, were asked to submit their views and proposals for a future sloop. Both were made quite independently (the Officers were serving on different Stations) and were remarkably similar in concept, only the figures for tonnage etc differing to any marked degree. Both authors considered that the anti-submarine capability was paramount, and they both proposed a ship carrying two twin 4.7in 40 degree mountings forward (these were proposed for the TRIBAL class ships but were still under design in 1932), an ahead throwing A/S weapon akin to the 3.5in stick bomb mortar then under test in H.M.S. TORRID and 40 depth charges (the standard outfit for a destroyer of the period was 12). The most startling proposals were that space aft should be left clear for aircraft stowage (one report recommended a catapult, the other implied a crane and an amphibian aircraft) and the strong recommendation that provision must be made for merchant ships in the convoy escorted to be fitted with a landing on deck to recover the aircraft. A truly unusual and prescient view by Officers not normally regarded as being in the forefront of advanced thinking! While the landing on deck was, certainly, little more than a pious hope at the time, and the aircraft concept would have required a considerable increase in tonnage and cost, it is worthy of note that the magnificent US Coast Guard cutters which were built very shortly afterwards did have provision for catapult launched float planes.

The pre-war LOWESTOFT at Plymouth prior to sailing for foreign service.

The Staff requirements for the GRIMSBY class are also worth examining: their endurance is specified as 5,000 miles at 10 knots but is also described as "the longest passage probable as an escort to a convoy, ie passage UK to Freetown, 3,500 miles plus a margin for operations". Again a significant pointer to Admiralty thought on a forthcoming conflict, and this before the elections that brought Hitler to power.

Subsequent sloop development up to 1944

The Staff Requirements for future sloop construction now underwent major change, reflected in the somewhat chequered design history of the next, BITTERN, class ships. Originally intended to be armed with four single 4.7in destroyer type mountings, ie bereft of any real HA capability, none was so completed. The lead ship, to have been named BITTERN, was completed for service as a new Admiralty

ENCHANTRESS on 9.3.36, in her early days as Admiralty Yacht. Although fitted out with the Board apartments aft, and painted in the Victorian colour scheme, she still mounts the midships 4.7in LA in Q position. This was later removed and an additional deckhouse fitted in lieu.

Yacht and mounted only her two forward and one midships gun initially, later reduced to the two forward mountings. Fitted aft with accommodation for the Board, she was named ENCHANTRESS to perpetuate that name, while completing with all the necessary internal arrangements to enable her to convert for wartime service as a fully armed vessel. The second ship built assumed the name BITTERN, and completed as the first unit with an all twin 4in gun armament; while the third ship, named STORK, completed as an unarmed survey ship although she later totally re-armed and emerged as a fully combatant vessel.

The BITTERN design mounting three twin 4in HA produced the most effective small AA vessel to date, but the Staff had to accept that there were two major faults in the ship as completed. The lack of a proper High Angle Control System (which was to bedevil the Royal Navy throughout and beyond the Second World War) rendered the ship less useful. Unfortunately, the defect could not be rectified for, incredibly, it proved quite impossible for British industry to design and build an effective system during the lifespan of the ship or her successors. Also, the inevitable motion of a small hull at sea affected AA efficiency; this could be reduced if not eliminated by the new fin stabilisers and the Admiralty took the drastic step of fitting these to new sloop and small AA ship construction. In the event the penalties, both financial and in loss of space internally, proved not worthwhile, but it was a bold and imaginative step on the Service's part. The class is also notable as marking the first step away from the system of naming sloops after the smaller ports and havens, henceforward bird names were to be used until the supply ran short during the war years, when traditional small ships' names of the 18th and early 19th century were brought back into use.

The Royal Indian Navy placed, via the Admiralty, orders for 4 ships based on a modification of the BITTERN design; the orders being given in two pairs which became, respectively, JUMNA, SUTLEJ, GODAVARI and NARBADA in that order. These ships are usually referred to as BLACK SWAN and Modified BLACK SWAN classes respectively; however official handbooks specifically refer to them as Modified BITTERN class.

The Indian NARBADA fresh from the builders prior to sailing East to her home base.

Three further ships followed, the EGRET class, which developed the AA capability by fitting a fourth twin 4in mounting aft where it finally displaced that monument to the sloops' origin, the minesweeping winch, which had still been fitted in all the preceding units. The class was also noted for another

National Maritime Museum Negative N22363
A rather drab PELICAN of the EGRET class lies at anchor. From the paintwork, lack of a jack and what appears to be a Red Ensign and builders flag at the main it is assumed that this is a trials photograph prior to acceptance by the Royal Navy. If so, there is yet a lot of work to be done!

renaming when one unit completed as AUCKLAND. It had been intended originally that she would serve on the New Zealand Station with WELLINGTON, hence the name. The international situation, with European tensions rising, did not make it wise to allow a modern unit to go to a remote Station, so she was retained in waters closer to probable action. Until 1941, the Royal Navy provided the ships for the New Zealand Station while making considerable use of New Zealand personnel; the Royal New Zealand Navy (as a separate Service) did not form until 1941.

The final sloop design, from which all wartime construction derived, was the well-known BLACK SWAN class. Ostensibly very similar to their predecessors, the class reverted to the three twin 4in HA main armament but with the addition of the new quadruple 2pdr mounting aft to provide a reasonable degree of close range AA defence. 4 ships were ordered to this design. The final Modified BLACK SWAN design, which concluded the sloop programme, consisted of 27 ships for the RN and two more for India. It must be noted that the 5 repeat ships WHIMBREL, WILD GOOSE, WOODCOCK, WOODPECKER and WREN, usually referred to as repeat BLACK SWAN class, are listed officially as of the Modified design. An additional 5 ships ordered of the Modified design were cancelled prior to construction commencing, so that the final tally of sloops constructed from 1926 over a period of twenty years was 71, 59 for the RN, 4 for the RAN and 8 for the RIN.

Had it not been for the disastrous inter-war years for the Canadian economy, the total would have been increased as the Admiralty had strongly recommended the original design to the Canadian Government when they sought replacements for the ageing HMCS PATRIOT and PATRICIAN. While finance precluded new construction at that time, and the proposal was not followed up, there is no doubt that such ships, and the possible HALCYON class also considered in the late 1930s, would have proved of great value to the RCN.

BLACK SWAN lying in the Mersey off Birkenhead in 1941. Note the early 286 "mattress" at the masthead, no other radar is fitted as yet. The quadruple 2 pdr is on the quarterdeck, which is also cluttered with the eight DC throwers required to handle the 14 charge DC pattern then in vogue.

The BLACK SWAN and Modified BLACK SWAN classes

The final design and its derivative (the later units of both classes were altered somewhat internally) were not quite as successful as their predecessors due to the age-old problem of pint pots and their contents. The basic sloop hull had been lengthened by 20% in the first twelve years of design, but even this increase in dimensions was insufficient to accommodate the increasing quantities of personnel, stores and equipment that wartime development and service imposed. In consequence, gross overcrowding of humans and commodities occurred and the hull structure became overloaded. In particular, an unusual problem manifested itself; with the removal of the minesweeping winch from the original design and the elimination of the twin 4in HA or quadruple 2 pdr which took its place, the ships were light aft with a consequent tendency to trim by the head. This situation was offset by the weight of the large depth charge armament, 100 charges being carried aft. However, it was not uncommon for these ships to expend their entire outfit in one cruise thus upsetting the trim at sea. In its turn, this magnified the stresses experienced forward, in way of the bridge structure, and aft at the break of the fo'c'sle, leading to cracked plating and hull distortion. In particular, Captain F J Walker (of 2nd Escort Group fame), was particularly outspoken about such defects in his ship, STARLING, referring to the ship as "shoddy, ill designed and of poor workmanship" and alleging that the class, and especially his own ship, was unseaworthy.

While there is no doubt that the increasing load factor had stressed the hulls to and beyond the limit, especially in heavy weather, it is probable that Captain Walker's exertions in hunting submarines were also equally detrimental. He was not noted for his qualities as a ship handler, but justly renowned for his determination to press to the limit when on operations. In fact he died from the effects of overstrain to which he had exposed himself in his famed career as the premier submarine hunter of the War. While alterations were made to the later ships of the programme, both by altering constructional detail and by increasing the beam, it is noteworthy that the lightly built CAPTAIN class frigates, operating at the same time and in the same area as Captain Walker when he made his complaints, did not suffer similar problems. Post-war service in the Far East, at times in typhoon conditions, did not reveal evidence of incipient damage.

While the early sloops had been designed for a dual escort and minesweeping role, and to be employed as cruising ships on distant Stations, the later designs maximised the anti-aircraft potential. It is ironic, therefore, that their principal wartime achievements, and the attendant publicity, were gained in the anti-submarine role. The well designed, long endurance hulls of the later classes (with an initial ability to absorb additional weight) made them excellent long range escorts, and later hunters, and they were principally employed as such from 1942 onward.

Wartime service

In the first months of the war while on passage to the UK from their overseas duty, the earlier, smaller, sloops demonstrated that they were quite capable of acting as North Atlantic escorts. Unfortunately, operational needs elsewhere limited that service and it was not until 1941 that they really began to be used as long distance escorts on the North Atlantic and Freetown routes. The modern ships, with a good HA armament, were widely used in the Norwegian campaign and on AA escort duties in UK coastal waters and the Mediterranean until increasing fighter protection became available.

In 1942, when the newest units began to come into service and the increasing submarine threat obliged every effort to be dedicated to convoy defence, it became possible to form some units into specific Escort Groups to be employed in the Support role. This did not imply a reversion to the pre-war, flawed, concept of the "Hunter" where ships were aimlessly to roam the seas looking for submarines. Rather it allowed them to be attached to convoys in addition to the normal escort, and therefore free of the limitations imposed on an escort by the convoy orders which laid down that "the safe and timely arrival of the convoy" was the prime duty. In simple terms, submarines existed to find and attack convoys, the escort existed to pass the convoy through the threat; if there were a second component (the Support Group) then once the submarines had found the convoy, that Group could remain and hunt them, to destruction. A submarine, once located and forced to dive was, providing the attacker could remain on the scene, almost inevitably sunk; in those pre-nuclear days a submarine could remain submerged for a finite period only, after which it either surfaced, or the crew asphyxiated. Once surfaced, it fell victim to the waiting ships. The system relied solely on the ability of the hunter to remain, a luxury not permitted to the escort. Captain Walker, in particular, perfected his Group's training to the extent that they rarely lost a submarine once located. For such Groups, a convoy became a honey pot which attracted the wasps; the Groups could be (and were) diverted from convoy to convoy as signal intelligence revealed which had been located so that arriving submarines found the executioners waiting. The ability to "sit on" targets was a philosophy totally in line with the offensive doctrines favoured by Walker, Macintyre, Gretton and the other successful anti-submarine commanders of the 1942/45 period. Walker was the epitome of that small band, his own ship STARLING took part in the sinking of no less than sixteen submarines in the last three years of the war.

Sloops were active in areas other than the North Atlantic, and anti-submarine operations of course; much of the anti-aircraft protection afforded during the Norwegian campaign, in the first two years in the Mediterranean, off the east coast of the UK and, in the final year in the Pacific, the protection of the Fleet Train off Okinawa and Japan itself, was provided by sloops. Nor must the activities of the RIN be overlooked; their sloops were employed initially in convoy escort duty in the Indian Ocean, and their lack of success is due to the paucity of targets which prevented them demonstrating their abilities. Once the campaign to re-occupy Burma got under way in 1944, these ships came into prominence as the "capital ships" of the coastal operations, penetrating well upstream in the rivers to provide fire support for the Army ashore, and a secure base for the small craft operating further up stream. Just how far "up the creek" a sloop could get is amply shown by the accompanying photograph.

This could almost be titled "Up the creek", and shows just how far up Burmese chaungs the sloops penetrated in support of small craft and the Army in the advance down the Burmese coast in 1945. Here FLAMINGO rests between operations, sampans alongside and ratings at ease on the quarterdeck.

Post war service

Sloops remained in service post war, although really unfitted for prevailing requirements. The Naval Staff had to accept that the hull size prevented them being refitted to modern AA and A/S standards (they were obliged to retain the by then antiquated twin 4in HA mounting, and could spare neither space nor weight to fit the SQUID or Mk10 A/S mortar), but they were still cost-effective ships in terms of manpower and finance. Hence the handsome BLACK SWAN design lived on post-1945 (together with a few of its immediate predecessors) in their traditional role as cruising ships on overseas Stations. SNIPE and SPARROW spent several years representing Britain in South American waters, including getting trapped in the ice when operating in the Far South. In the Far East it will be a very long time before the involvement of the AMETHYST in the Yangtse incident is forgotten. However dubious the political moves which placed the ship in that position, her actions and the conduct of her ship's company will long be remembered. At home, for a number of years post-war, a sloop remained as the Senior Officer's ship of the Fishery Protection Squadron operating both in British waters and in the far north supervising distant water fishing fleets.

Even in the "peacetime" years sloops were involved in action such as the Israeli/Arab war when CRANE was mistaken for an Egyptian warship and was attacked by Israeli aircraft. Luck was with the ship that day, the aircraft were armed for ground targets with graze fuzed rockets so that the strikes did not penetrate the hull and the ship escaped serious damage while shooting down one of her attackers.

I. L. Buxton

SNIPE, outer, and PHEASANT laid up at Barry 11.8.56 with WIGTOWN BAY astern.

The end of service

The inevitable end of the post-war sloops was the Reserve Fleet and then the final passage to the breaker's yard. For most, this had taken place in the late 1950s and early 1960s, but some lingered on in other roles and service. The oldest existing sloop, WELLINGTON went early, sold to the Honourable Company of Master Mariners for use as their Livery Hall. A fitting end for a ship with a gallant career, she was gutted internally and fitted for her new function, which she still carries out, being moored in the Thames ahead of the 1916 CHRYSANTHEMUM so that there are two sloops still extant in London's river. How many Londoners and visitors realise that the smart, white painted, buff funnelled WELLINGTON against the Embankment fought in the Battle of the Atlantic I do not know. Of the Indian ships, the KISTNA survived the longest, as a training ship.

Overseas service

After 1945, many nations were obliged to rebuild their navies with purchases/loans from the major Powers, in some cases indeed to create a Fleet. While much of the tonnage involved came from American sources, Britain did not lag far behind, and sloops formed a surprisingly large proportion, indeed one of the earlier sales remains the sole surviving representative of the type. WHIMBREL was sold to Egypt as a unit of the re-formed post-war fleet, and became EL MALEK FAROUK, later renamed TARIQ by the Republic. As such, she still remains in service, though only as a static depot ship in recent years, an example of the longevity of these ships, and a memory to all that served in the Royal Navy. Currently efforts are being made to raise funds to purchase her and return her to a berth in the United Kingdom.

Other overseas service was in the newly formed Bundesmarine when West Germany was permitted to form its own Armed Forces. ACTAEON, FLAMINGO, HART and MERMAID were sold and commissioned as HIPPER, GRAF SPEE, SCHEER and SCHARNHORST respectively, for use as training ships, a role in which they served for a number of years. Although the other three were eventually paid off as redundant together with other purchased British hulls, SCHARNHORST lived on until 1990 in service as a damage control hulk before finally being broken up in Belgium.

courtesy Bundesmarine
Taken after conversion for German service, ACTAEON as HIPPER in 1959. She is still recognisable as a British sloop, the fitting of German close range weapons amidships, a funnel cap and the extra classrooms aft do not greatly alter her silhouette. She retains her British radar, but has lost the 291 from the masthead.

Sloop conversions

The design process whereby sloops were both minesweepers and patrol vessels has already been discussed; however there were other dual roles that the sloop was obliged to undertake, involving varying degrees of conversion.

Duty on the China Station based on Hong Kong involved the early sloops in the defence plans for the Colony which, amongst other matters, required the laying of defensive minefields. To provide the necessary capability, plans were made for the sloops to be capable of a limited minelay effort. This involved the removal of top weight, ie the minesweeping gear (excluding the winch), and X gun, and rigging mining rails aft etc, the necessary rails and gear being held locally ashore. About 24 hours was required to carry out this work, after which the ships could embark and lay, manually, 32 buoyant mines.

The M/S to A/S and sloop to M/L conversion could, almost, be undertaken without Dockyard assistance, but there were other roles required of the sloops that involved physical alteration to structure, internal and external.

Overseas, the Commander-in-Chief usually wore his flag in a cruiser, but many of the ports which he might wish to visit (such visits were analagous to the State Visit of the present day) could only be made in a smaller vessel with lesser draught. There was, therefore, a real need for a small ship with the facilities appropriate to a Senior Officer, particularly on the China, Africa and Mediterranean Stations. To this end sloops were altered by the removal (or omission on building) of the after gun and the installation of a cabin block in lieu which provided living accommodation for the C-in-C, his wife, retinue and staff. The ship was otherwise fitted out for war, the magazines, for example, being fitted for shell though used as baggage stores, all wiring etc for guns and gunnery control being fitted. The actual cabin accommodation could be removed - dockyard assistance was needed to cut it away - but no actual structural alteration to the ship was involved.

The conversion of BITTERN into the Admiralty Yacht ENCHANTRESS involved somewhat more work, and the extra after-structure remained in place during her war career. This was, however, probably due to lack of a dockyard work slot than any other reason; the ship completed otherwise to an operational condition. She is noteworthy as the last Royal Navy ship to wear the old Victorian livery of black hull, white superstructure and buff funnel; and very smart she looked as the photographs show.

The final conversions, made during building or at a later stage in the ship's career, involved the use of sloops as Survey Vessels. Again (if newly built) the ship was completed "for but not with" armament etc, but major work was needed to add extra boats, equipment and chartrooms - in those days the Officers who carried out the survey actually drew the Master Chart onboard daily. If converted after completion as a sloop, the combatant fittings inboard (wiring, magazines fittings etc) remained, where possible. Examples of sloops so converted, up to the latest building pre war, are shown in this work. In all cases the ships went to war, either in the form to which they had been converted, being altered later to conform to their sisters, or were completed to sloop condition before the conflict.

The Indian JUMNA post war as Leader of 12 FS, date unknown but possibly approaching Portsmouth for the 1953 Review.

Armament changes

Individual changes of armament are listed elsewhere but an overall review of developments is of value. In the main, the principal armament, the 4in, was retained during the ships' lives although LA weapons were replaced by HA, and in a few cases, a twin 4in HA mounting was provided in lieu of the single, while the twin 4in HA mounting became standard for the later classes.

Close range armament had to await war time developments, the 0.5in being retained in many sloops until late in the war. Progressively, this mounting was replaced with the 20mm Oerlikon which in its turn was supplemented and then supplanted by the 40mm Bofors. Only the most modern designs were capable of fitting the quad 2pdr mounting that was such a feature of British flotilla craft.

It was in the anti-submarine field that the greatest changes occurred. From the beginning it was intended that sloops should possess such a capability, but on a very small scale. ASDIC was provided for in all ships, but frequently not fitted at first. This was, in fact, a blessing as, when sets were installed in the late 1930s in existing ships, they received a later mark of equipment than would otherwise have been the case.

The standard A/S weapon was, of course, the depth charge. The early sloops carried 4 depth charges, fitted in chutes for over stern launching! Frequently, these chutes were not fitted but carried onboard in store. From ABERDEEN and FLEETWOOD onwards two depth charge throwers and sets of depth charge rails were fitted, with an increase in stowage to 40 charges, this suite being applied to the earlier vessels as time permitted. The throwers were, of course, the old Mk II that required a separate carrier that was expended with each depth charge.

WREN in a typical Clyde portrait. An early photograph, she still has the eight DCT armament and only a light 20mm, four single weapons in the bridge and amidships. 291 has not yet been fitted.

With the BLACK SWAN design, major changes were made early in the war. The number of throwers was increased to 8 to permit the use of the later discredited 14 charge pattern, the Mk IV thrower with its integral carrier was provided, and the depth charge outfit increased to 110. The subsequent recognition that the 14 charge pattern set for three depths was a flawed concept, owing to it counter-mining its own charges, caused a reversion to a 10 charge pattern set at two depths, permitting a reduction to four throwers, easing the congestion on deck aft. Thereafter four throwers and two rails became the standard for all anti-submarine craft, and ships with eight throwers were reduced to 4. Ship size dictated the outfit of charges, the larger, more modern sloops carrying 110, the older ships from 60 to 90.

In 1942 the advent of "Hedgehog" brought a considerable, potential, increase in submarine killing capability, unfortunately not fully achieved for almost 18 months. The Hedgehog mounting was fitted on the foc's'le, sometimes offset to one side to provide space for the ready use ammunition locker. In the twin 4in ships with two mountings forward, there was insufficient space, but it was not acceptable to remove one of the mountings. A new Hedgehog had therefore to be devised, and became known as the "split Hedgehog". Quite simply, the mounting was divided in two along its centreline, the two halves shipped on B gun deck either side of the 4in, being cross connected by tie bars to form an integral structure; its operation was otherwise identical to the usual version. Magazine stowage was for a standard 144 bombs, providing six full salvos of 24.

Locally entered personnel

Earlier in this narrative, mention has been made of provision for "native" crew members. It should be understood that from the mid-Victorian times, the RN made considerable use of locally recruited ratings, known in the Service as "Locally entered personnel" or "LEPs".

Five principal sources of such ratings have existed, Malta, Aden, West Africa, Goa and Hong Kong. The last named recruiting centre still (1993) exists albeit in a much reduced role.

Such recruitment was intended to provide additional, or more suitable, manpower for onboard service in foreign waters. Maltese ratings were largely employed as cooks and stewards, as were Chinese; however both nationalities also supplied seamen and engine room personnel for service locally in the Colony and in ships on the appropriate Station.

Goanese ratings, recruited only in small numbers over a limited period, were used as stewards and messmen, while the use of Kroomen and Seedie "boys", the titles applied to West African and Somali ratings respectively, was restricted to general provision of labour in climates inimical to European ratings.

Only Chinese and Maltese ratings survived into the post-1945 era, the Maltese recruitment ceased on the independence of Malta and the ratings were discharged save for those who opted to transfer to normal engagements in the RN, which some did. In 1993 only a small number of Chinese LEPs remain, onboard as cooks and stewards and ashore in Hong Kong as general service ratings employed in TAMAR, the shore establishment in the Colony.

BRIDGEWATER class

Name	Builder	Laid down	Launched	Completed
BRIDGEWATER	H Leslie	6.2.28	14.9.28	14.3.29
SANDWICH	H Leslie	9.2.28	29.9.28	23.3.29

Displacement 1,045 tons standard. Dimensions length 266ft 4in oa, beam 34ft, draught 11ft 5in aft at full load.

Machinery two shaft, geared turbines, designed SHP 2,000 = 16.5 kts. BRIDGEWATER made 17.24 knots on trials, SANDWICH 17.27. Oil 282 tons, consumption 0.54 tons per hour at 10 knots.

Initial armament one 4in HA forward, one 4in QF LA aft, two 3pdr saluting guns. Two chutes with capacity for 4 DCs carried.

Changes during service
During refits in 1938, both ships replaced the after 4in with a second 4in HA, removed the 3pdr guns and mounted two quadruple 0.5in mg.

In 5.42 BRIDGEWATER, and in 7.42 SANDWICH, replaced the quadruple 0.5in with two single 20mm Oerlikon.

In 5.43 BRIDGEWATER, and in 10.43 SANDWICH, the 20mm were increased to four guns, and Hedgehog was fitted.

Radar

BRIDGEWATER fitted 271 in 5.42, and HF/DF in 5.43.

SANDWICH fitted HF/DF in 7.42, 271 in 9.42, replaced by 271Q plus 291 in 9.43.

BRIDGEWATER lying in Plymouth Sound after completion, the first sloop built post 1920. Note the basic superstructure and simple fittings; two single 4in fore and aft are the sole armament, the solid bulwark sweep deck houses the large minesweeping winch, M/S gear and davits right aft. She has a white hull and buff superstructure in preparation for foreign service.

HMS BRIDGEWATER L 01, U 01

BRIDGEWATER was one of the first two ships built to replace the ageing HUNT and FLOWER class of 1916. Launched without a formal ceremony, she completed in 3.29 with a Devonport crew for service on the China Station, where she and her sister SANDWICH replaced BLUEBELL and FOXGLOVE, all ships meeting at Aden to transfer Chinese ratings.

BRIDGEWATER remained on the China Station until 1.35 when she refitted in South Africa prior to joining the Africa Station, later titled the South Atlantic Station. Here she remained until her return to Devonport 1.9.36 on completing her third commission, two replacement crews having been sent out from the UK in the seven years of service.

After a month's refit, BRIDGEWATER recommissioned and returned to Simonstown for further service. Refit at Simonstown, when she fitted her second HA gun, was interrupted by the Munich crisis and her despatch to Freetown. She returned to complete refit, and remained on station, arriving home 13.4.39 at Devonport to refit and recommission.

BRIDGEWATER sailed 17.5.39 for passage to Simonstown via a long cruise down the West coast of Africa, arriving in 8.39 to be sent back to Freetown for war service. She stayed there briefly and then returned to the Cape until 1.40 when she went back to Freetown, now the main base for the South Atlantic Station.

Lack of docking facilities there, and the rapid fouling of hulls in those waters, occasioned her return to the Cape for docking in 6.40. Further visits were to come in 2.41, and 1.42. BRIDGEWATER completed her first visit in time to take part in the abortive Dakar attack, Operation Menace, after which she resumed convoy duties from Freetown, one major operation being her escort of WS 16 from Freetown to the Cape in 3.42. In 7.42 she took SL 114 to the UK, and went to refit on the Tyne, her first major refit since leaving the building berth.

BRIDGEWATER in her final guise. Hedgehog can be seen forward of A gun, there are 20mm in the bridge wings. The 271 lantern is at the back of the bridge, 291 at the masthead. Right aft, a tall mast carries the aerial of HF/DF.

On completion, the now ageing ship returned to Freetown with OS 42 arriving 9.10.42, just one month before Operation Torch, the North African invasion. BRIDGEWATER continued to serve at Freetown until late 1943, but had a welcome change of scene when she escorted the damaged cruiser PHOEBE to Trinidad en route to the USA, and returned with convoy TF 1 to Freetown.

Her subsequent duties involved the southern end of the Freetown/Gibraltar convoy series RS/SR, and local West African convoys, prior to her return to the UK with convoy SL 136, arriving Liverpool 25.9.43, to refit at Southampton and then to join 40 EG. However, the worn-out hull did not permit this, and she had to be re-allocated to the less arduous duty of submarine escort with the 3rd Submarine Flotilla at Holy Loch where she acted as an escort and target ship. She remained in this role for the rest of the war, paying off to Category C reserve at Ardrossan in 7.45.

Used for static bomb trials, the hulk was transferred to BISCO 22.5.47, allocated to Howells and broken up at Gelleswick Bay.

HMS SANDWICH L 12, U 12

On completion, SANDWICH proceeded with BRIDGEWATER to China having commissioned with a Portsmouth crew, embarking her Chinese ratings at Aden from the ship she was relieving, a commission that continued until her relief crew arrived at Hong Kong to re-commission her 19.10.31. This crew continued the China Station duty until also relieved at Hong Kong 2.4.34. She remained on station even longer than her sister, for her re-arming with 4in HA took place at Hong Kong 6-10.38 and she recommissioned 21.3.39 with yet another crew sent out from the UK.

War found SANDWICH still at Hong Kong, and she was employed cruising in the Tsushima Strait off Japan to intercept any German merchant vessels that might be in that area. In 10.39, a brief period at Hong Kong completed her for A/S work and after one more patrol she left Hong Kong for the UK, via Gibraltar where she picked up her first convoy, escorting HG 11 to Devonport in time to give Christmas leave, her first visit home since completion almost twelve years earlier.

A somewhat obscured photo of SANDWICH leaving harbour.

On completion of refit in 1.40 she escorted in the South West Approaches followed by a complete passage to Gibraltar with OGF 21, returning to Liverpool with HG 24, and was thereafter based there, operating Gibraltar routed convoys until 17.5 when she turned her attention to the North Atlantic route with OB 154 on 24.5.

In 7.40 came a change of scene when SANDWICH moved to Rosyth to escort the OA convoys around Scotland and into the Atlantic, a duty that continued until 10.40 when she took OA 223 well into the Atlantic, thence to Sydney CB to return to Liverpool with SC 8. Thereafter she was Liverpool based for a short period, during which time she rescued survivors from *KING IDWAL*, and *ANTEN*, prior to a long refit at Tilbury 12.40-4.41.

After refit SANDWICH did one trip to Gibraltar with OG 59 and HG 61, and then returned to the North Atlantic until 8.41. After this she was employed on the slightly less boisterous Freetown route commencing with OS 2 in 8.41, duty that continued until 1.42 when she went for refit at Belfast and to change her crew.

Returning after work up to Freetown, this remained her outward destination until 10.42. During this time, while escorting OS 35, she participated in the sinking of U 213 on 31.7. In 10.42 she had a brief period in Belfast to fit her 20mm, and then sailed 26.10.42 with KMS 2, the second stores convoy of

SANDWICH does not appear to have been blessed with good photographs, this Coastal Command picture taken 7.42 hardly does her justice but does show the changes made since the previous photograph.

Operation Torch, the North African invasion, following which duty she remained based at Gibraltar for the follow-up operations.

SANDWICH returned for a much needed refit on the Tyne from 2-7.43 and then proceeded to Freetown, operating between there and Gibraltar with the SL/OS convoys until 6.44, when she was sent from Gibraltar to Brindisi, where she appears to have stayed until 11.44. It is assumed this was for refit in one of the Italian dockyards then under British control; further work was required at Gibraltar that lasted until 1.45.

Presumably the ship was now so worn-out as to be beyond effective repair, for in 1.45 she reduced to Care and Maintenance at Gibraltar, and a month later she was moved to Bizerta to lay up in the anchorage there. There she remained, until sold (locally) to an unknown buyer on 8.1.46 for the meagre sum of £3,050.

HASTINGS class

Name	Builder	Laid down	Launched	Completed
FOLKESTONE	S Hunter/ H Leslie	21.5.29	12.2.30	25.6.30
HASTINGS	Devonport/ V Armstrong	29.7.29	10.4.30	26.11.30
PENZANCE	Devonport/ V Armstrong	29.7.29	10.4.30	15.1.31
SCARBOROUGH	S Hunter/ H Leslie	28.5.29	14.3.30	31.7.30

Displacement 1,025 tons standard. Dimensions length 266ft 4in oa, beam 34ft 1in, draught 11ft 3in aft at full load.

Machinery two shaft, geared turbines, designed SHP 2,000 = 16.5 kts. Oil fuel 280 tons, consumption 0.6 tons per hour at 10 kts.

Initial armament one 4in HA forward, one 4in LA aft, two 3pdr saluting guns. Two chutes carried with capacity for 4 DCs.

Changes during service

HASTINGS and PENZANCE both shipped a second 4in HA, the former in 1938, the latter in 1939. All four ships fitted their DCTs and rails and increased DC stowage to 40 on the outbreak of war.

PENZANCE had no further, known, alterations prior to her loss.

FOLKESTONE was disarmed as a Survey Vessel in 5.39, and re-armed in 12.39 with a single 4in LA and DC equipment. She did not ship her 4in HA until 7.41, at which time she also acquired two single 20mm to supplement the two quadruple 0.5in mg mounted earlier, the 0.5in were later removed and two further 20mm and Hedgehog added in 7.42.

HASTINGS acquired two quadruple 0.5in mg shortly after 9.39, supplemented by two 20mm in 7.41. Hedgehog was added in 10.42.

SCARBOROUGH was disarmed in early 1939 as a survey ship, and re-armed with a single 4in LA and a 12pdr and two quadruple 0.5in mg by 12.39. This armament was updated in 7.42 when two 20mm Oerlikon were fitted, increased to four in 10.42. In 6.43 armament was further increased when Hedgehog was fitted.

Radar

In all except PENZANCE, 286M radar was fitted in 1941, replaced by 271 progressively in 1942, HF/DF and Hedgehog were also fitted in 1943, with 291 also in FOLKESTONE at least.

HMS FOLKESTONE L 22, U 22

FOLKESTONE commissioned with a Portsmouth crew on completion for service in the Persian Gulf, arriving there in 8.30. She remained on Station, except for a drydocking at Bombay 1.31, until 8.31 when she went to Colombo to refit and exchange crews with FOWEY. Having done this, and completed refit 4.11.31, the ship then became part of the China Station and operated from Hong Kong. She recommissioned twice while on station, at Hong Kong 2.4.34 and at Singapore 8.10.36, remaining on the China coast until 5.39 when she was taken in hand at Hong Kong and converted to an unarmed Survey Vessel intended for service in New Zealand; however FOLKESTONE never thus served due to the outbreak of war. Part of her crew manned a minesweeper from reserve and in 12.39 she was fitted out as an A/S vessel with a single 4in LA, and ordered home for service.

The new FOLKESTONE lying in Spithead 20.10.30 in her overseas colours. Note the prominent M/S gear and the 1916 vintage 4in LA aft in its destroyer type shield. Close range and A/S armament is still non-existent.

FOLKESTONE arrived at Portsmouth 14.2.40 to give leave and refit, then joined Western Approaches Command at Liverpool for North Atlantic work. She collided with, and sank, RIVER HUMBER 4.6.40, repairing at Cardiff and had further Liverpool refits in 12.40 and 4-5.41. She had a minor collision with an unknown vessel 11.41, and after repair she became part of the newly formed 42 EG operating to and from Freetown.

FOLKESTONE had further refits at Cardiff 4-6.42, Londonderry 10-11.42 and at Grimsby 4-5.43 prior to joining the West Africa Command, based at Freetown. She returned to Liverpool 9.43 for a 6 week refit and on 13.11.43 suffered a boiler explosion. However this must have been minor for it did not delay her return to Freetown. In 3.44, while still so based, she became part of 56 EG, but returned home in 9.44 to lay up at Milford Haven, then being beyond economic repair. She remained in reserve until utilised for bomb trials post war, passing to BISCO 22.5.47 and being broken up at Gelleswick Bay.

FOLKESTONE in war paint, literally. She now has a 4in HA aft, 271 and 291 radar, HF/DF on its tall mast aft and 20mm in the bridge wings. The sweep deck bulwarks have been cut back to clear the DCTs now fitted, and the M/S gear is long since landed. Photographed early in 1942, possibly at Londonderry, certainly prior to her mid 1942 refit when Hedgehog was fitted.

HASTINGS at the Coronation Review 1937. As a Senior Officer's ship she wears the black funnel top of a leader. Forward she has only an old destroyer 4in LA dating from 1916, otherwise she is fitted only with her minesweeping gear. As a Leader, she does not have her pennants painted up.

HMS HASTINGS L 27, U 27

HASTINGS commissioned with a Devonport crew for service in the Persian Gulf, which lasted from 1.31 to 1.32 when she transferred to the Red Sea. Shortly before leaving the Gulf she was involved in the salvage of *BARODA* and *ORMONDE*, both in 1.32.

Serving in the Red Sea meant refits at Malta rather than Bombay, and visits there occurred in 2.32, 8.32, 3.33 and 9.33. The ship re-commissioned at Malta 3.34 for further Red Sea service which came to an abrupt stop on 11.6.34 when she grounded heavily, not being refloated until 6.9.34. She was then towed to Suez for temporary repairs. On completion, the cruiser DURBAN towed the damaged ship to Malta where she was taken in hand for permanent repair. This lasted from arrival in 11.34 to 4.37, work being suspended during the Abyssinian crisis 4-9.35.

On completion at Malta, HASTINGS commissioned with a temporary crew and was steamed home, to re-commission at Devonport 5.37 for the Fishery Protection Squadron as Senior Officer. As a change from these duties, the ship fitted her minelaying gear at Portsmouth 8-9.37 returning to her FP duties afterwards. Further refits were at Devonport 12.37 and again 4-8.38 when she re-armed, with a final pre-war refit also at Devonport 7-9.39 during which her minelaying gear was removed and A/S gear fitted.

A more warlike HASTINGS, now with two single 4in HA, 20mm in the bridge wings, a full A/S outfit after landing all the M/S gear including the winch, 271 and 291 radar and HF/DF. Oddly, she retains her quadruple 0.5in in the after bandstands. Hedgehog has been fitted right forward just abaft the capstan.

HASTINGS's first wartime duty was with the Rosyth Escort Force on the East Coast, during which she was in collision with *BRADMAN* 1.1.40 and *LIMESLADE* 1.12.40. Transferred to Western Approaches Command at Liverpool 7.41 she joined 44 EG 12.41 and 40 EG 2.42 for service on the Freetown route. HASTINGS also operated in the Support Group role 4-8.43 prior to joining 39 EG later in 8.43.

Presumably due to defects and general age, HASTINGS paid off at Belfast 19.11.43 and went into reserve at Hartlepool 2.44. She was taken in hand 5.44 and refitted between then and 9.44 for duty as a submarine escort, being attached to the 3rd Submarine Flotilla on completion, serving at Holy Loch until 2.46. Paid off there 16.2.46, HASTINGS passed to BISCO 2.4.46 and arrived at Troon 10.4.46 to be broken up.

PENZANCE in 5.39. She has re-armed with her second HA and has a new bridge, otherwise she remains as built with the quadruple 0.5in mg her only close range armament. She still carries full M/S gear.

HMS PENZANCE L 28, U 28

PENZANCE completed for Persian Gulf service with a Devonport crew and served on station until 10.31, when she exchanged crews with SHOREHAM and transferred to the Red Sea shortly afterwards. She refitted at Malta in 5.32, and carried the Emperor of Abyssinia, Haile Selassie, on a State Visit to Aden and Djibouti. Her first period of service finished at Malta 6.33 when she both refitted and re-commissioned, again for duty in the Red Sea.

PENZANCE was closely involved in the salvage operations following the stranding of HASTINGS, standing by the stricken ship for weeks on end and shuttling personnel, gear and stores to and from the nearest port. Following this work, she re-commissioned at Aden 29.11.35 and joined the Africa Station after a refit at Simonstown from 8.36-1.37, remaining there until returning to Chatham in 5.38.

On completing refit and re-commission, PENZANCE commissioned with a Chatham crew to relieve LUPIN in the Fishery Protection Squadron. However these orders were altered to retain a more modern ship in Home waters, so PENZANCE exchanged crews with PELICAN and proceeded to the America & West Indies Station instead, arriving at Bermuda in 6.39.

On the outbreak of war PENZANCE was based at Trinidad on contraband control and patrol work until 3.40 when she moved to Bermuda to escort the newly started BHX convoys taking ships north to join the HX convoy route to Britain. She acted as local escort for these convoys until ordered home in 7.40, when she went to Sydney CB and took charge of the first SC convoy from that port, SC 1, sailing 15.8.40. Unfortunately, she was lost 24.8.40 when torpedoed by U 37 while still with the convoy.

HMS SCARBOROUGH L 25, U 25

SCARBOROUGH completed with a Chatham crew and was intended to serve with the 1st Minesweeper Flotilla in Home Waters. However these orders were changed, and the ship proceeded to the America & West Indies Station, arriving at Bermuda 2.12.31 and serving on the East Coast of N and S America until returning to Sheerness to refit and re-commission 31.5.33 for further service on the original Station.

A superb photograph of SCARBOROUGH dis-armed and fitted out as a survey ship in 1938. Note the boom forward for the deep sea sounding gear, the increased number (and size) of boats, and the large chartroom aft in lieu of X gun and extending over the sweep deck. This chartroom was where the Surveying Officers prepared the original, master, chart on a daily basis.

SCARBOROUGH returned to Sheerness 8.38 to refit as an unarmed Survey Vessel for the East Indies and arrived at Colombo in 5.39. The imminence of war curtailed her activities, and she re-armed as an A/S escort 8.11.39 and returned to the UK arriving at Devonport 1.40 to refit and join 1st Escort Vessel Division at Liverpool 2.40 for North Atlantic work. This Group was later titled "Liverpool Sloops" and it was as part of this Group that SCARBOROUGH sank U 76 on 5.4.41 while with convoy SC 26.

Refit at Liverpool 8-9.41 preceded attachment to 43 EG for the UK/Freetown route, during which she was in collision with BRADFORD 18.4.42. Refit at Liverpool 7-10.42 preceded Operation Torch and later attachment to Western Mediterranean Command based at Gibraltar until 2.43, when she returned to Londonderry and 39 EG.

Transferred to 15 EG in 1.44, SCARBOROUGH took part in Operation Neptune and then laid up in reserve at Hartlepool 3.8.44, being worn out and beyond refit. She passed to BISCO 3.6.49 and arrived at Thornaby on Tees 3.7.49 to be broken up.

SCARBOROUGH in 8.43 in her wartime appearance as re-armed for escort duty. She now has a more modern 4in HA mounting forward, with Hedgehog abaft it to starboard. The new bridge has 271, and the foremast carries 291. There are single 20mm in the bridge wings and right at the after end of the shelterdeck. HF/DF is on the light lattice mast aft, and there is a full A/S fit of 4 DCT, reload racks and rails.

RIN HASTINGS class

Name	Builder	Laid down	Launched	Completed
HINDUSTAN	S Hunter/H Leslie	4.9.29	12.5.30	10.10.30

Displacement 1,190 tons standard. Dimensions length 296ft 4in, beam 35ft, draught 11ft 6in at full load.

Machinery two shaft, geared turbines, designed SHP 2,000 = 16 kts. Oil fuel 339 tons, consumption 0.6 tons per hour at 10 kts.
Initial armament two single 4in LA, four 3pdrs.

Changes during service
Other than the addition of 2 single 20mm in, probably, 1942; and the reported addition of a quadruple 0.5in mg earlier in the war, there are no reported additions to armament. Possibly, the 3pdr saluting guns may have been retained, certainly until the fitting of 20mm.

ASDIC was fitted in the latter part of 1940, and there are reports that Hedgehog was also mounted in 1943 but this is very dubious and is not supported by any known photographic evidence.

National Maritime Museum Negative N6095
India's HINDUSTAN clearly shows her origins in this pre-war photograph. She was little changed throughout her career.

HMIS HINDUSTAN L 80, U 80, F 80

HINDUSTAN served principally as a training ship in Indian and adjacent waters under the control of the Royal Indian Marine, later the Royal Indian Navy, until 1939. On the outbreak of war, in common with all Indian warships, she came under the operational control of the Royal Navy and was to all intents and purposes integrated as an HM Ship.

Initially HINDUSTAN served in the Persian Gulf area, one of her duties being the escort of the gunboats GNAT and COCKCHAFER on passage Bombay to Basra. A refit at Bombay in 9.40 fitted A/S gear, presumably ASDIC, and the ship returned to duty at Aden and in the Red Sea until 6.41. A further Bombay refit was followed by duty again at Aden until 12.41 when HINDUSTAN returned to Indian waters on the outbreak of war with Japan, going in early 1942 to Rangoon to assist in the evacuation there in 3.42. Thereafter the ship was based principally at Bombay and engaged on convoy duties between that port and Ceylon until late 1943, when she moved to the east coast of India and operated as far north as Chittagong.

Post-war the ship was disarmed and altered for duty as a survey vessel and training ship. On the Partition in 1947 she passed to the newly formed Royal Pakistan Navy as KARSAZ. Unfortunately, early records of that Service are scanty, and no further information except her reported scrapping in 1951 is available.

SHOREHAM class

Name	Builder	Laid down	Launched	Completed
BIDEFORD	Devonport/White	10.6.30	1.4.31	27.11.31
FOWEY	Devonport/White	24.3.30	4.11.30	11.9.31
ROCHESTER	Chatham/White	24.11.30	16.7.31	31.3.32
SHOREHAM	Chatham/White	19.12.29	22.11.30	2.11.31

Displacement 1,105 tons standard. Dimensions length 281ft 4in, beam 35ft, draught 10ft 4in at full load (Note: SHOREHAM is reported as 11ft 6in, the other three reached this figure in 1940).
Machinery two shaft, geared turbines, designed SHP 2,000 = 16.5 kts. Oil fuel 290 tons, consumption 0.54 tons per hour at 10 kts.
Initial armament one 4in LA, four 3pdr.

Changes during service

During 1938, all ships exchanged their 4in LA for a HA mounting and added a quadruple 0.5inch mg.

BIDEFORD acquired two single 20mm in 3.41, increased to four in 3.42.

FOWEY shipped an additional quadruple 0.5in mg at an unknown date, added two single 20mm, probably in 7.41, and in 5.43 added four further 20mm, landed the 0.5in mg, and added Hedgehog.

ROCHESTER added a second quadruple 0.5in mg early in the war, and then followed FOWEY by increasing the fit to two, later four, 20mm and Hedgehog. When refitted for service as a training ship, armament was reduced to seven single 20mm initially, later totally disarmed.

SHOREHAM had the greatest changes, progressively adding a second quadruple 0.5in, two increased to three single 20mm, then a single 2pdr pompom and, finally, in 7.45 when she returned to the Gulf, a 3pdr saluting gun.

Radar

All ships progressively fitted radar and amended the equipment,

BIDEFORD fitted 286, replaced later by 291. HF/DF was fitted in 6.42 and 271 in 3.43.

FOWEY fitted 286, replaced by 291 in 1941, added HF/DF 7.42 and 271 in 11.42.

ROCHESTER is noted as fitting 271 and HF/DF in 10.41, probably 286 had been fitted earlier, it was later replaced by 291.

SHOREHAM is noted as fitting HF/DF, 271 and 291 in 1.43, later than most and an indication of her Eastern service.

WSPL

BIDEFORD in her original form. Note the multitude of shrouds, dispensed with once tripod or lattice masts were fitted. The M/S davits aft are prominent, while the 4in LA aft displays its S class destroyer origins by its half shield.

HMS BIDEFORD L 43, U 43

BIDEFORD commissioned with a Devonport crew and proceeded to the Persian Gulf, exchanging crews with HASTINGS on arrival in 1.32. Thereafter she remained in the Gulf, twice assisting *ORMONDE* when that ship grounded in 1.32 and 3.32. BIDEFORD went to Bombay in 11.32 to re-commission with a new crew sent out from Devonport in the cruiser HAWKINS, returned briefly to the Gulf and then went to Colombo for refit arriving on Christmas Eve and remaining at Colombo until 17.2.33, a welcome break for her crew no doubt.

Returning to the Gulf, BIDEFORD remained on station until a further refit at Colombo 3-4.34, after which it was back to Gulf duty until a new crew arrived and re-commissioned the ship at Basra 6.11.34.

Despite this change, BIDEFORD remained in the Gulf until 7.36 when she arrived at Colombo for refit lasting until 15.8.36, then back on station with a further fresh crew joining, again at Basra, 17.11.36. The ship's further service ended with a two month refit at Bombay 9-11.37 following which she returned to the Gulf prior to going to Malta in 8.38 for a major refit and re-arming.

BIDEFORD's lengthy stay at Malta - she did not complete until late 12.38 - concluded with her embarking a passage crew and returning to the Gulf until 5.39 when she finally departed for the China Station, arriving at Singapore 21.6.39 where she exchanged crews with GRIMSBY and went onward to Hong Kong.

BIDEFORD was still on the China Station when war broke out, and remained there until 12.39 when she returned home for refit at Devonport, escorting convoy HGF 14 from Gibraltar on the final part of her passage.

Completing refit 2.40, BIDEFORD joined 1st Escort Vessel Division of Western Approaches Command and escorted Gibraltar bound convoys until sent to Dunkirk as part of Operation Dynamo, during which she was bombed 29.5.40. The damage incurred required repairs lasting until 4.41 when, after work up at Tobermory, BIDEFORD joined the Sloop Division of her old Command and continued to escort convoys on the Gibraltar and Freetown routes until 2.42.

An early wartime view of BIDEFORD. Note the cluttered quarterdeck, still with its minesweeping winch although all M/S gear has been discarded including the davits right aft. The 286 "mattress" can be clearly seen at the masthead.

A refit at Hartlepool until 4.42 concluded with a slight collision with *PETROPHALT* which, however, did not delay her return to the South Atlantic convoy routes where she was to serve until 1.43 when, after Operation Torch and a collision with the Canadian corvette LOUISBURG, she went to Avonmouth for refit 1-4.43. On completion and work up, BIDEFORD went to North Atlantic escort duty, both as a convoy escort, in Support Group duties and on Biscay A/S patrols in 7-8.43. During this last activity, she was near missed by a glider bomb on 25.8 and had to repair damage at Londonderry throughout 9.43.

In 10.43 BIDEFORD went to 41 EG and operated on the Freetown route, passing with that Group to the Mediterranean Fleet in 1.44 for duty on convoy routes there. Transferred to 50 EG 4.44, BIDEFORD continued in the Mediterranean until 1.45 when she came home to refit at Devonport during 1.45 and to join 41 EG, now Devonport based, on completion. BIDEFORD saw out her war service with this Group, escorting her final convoy (MKS 100) in mid 5.45, following which she de-stored at Cardiff and was laid up at Milford Haven 8.6.45. The old ship remained there until passed to BISCO 14.7.49 following which she was broken up locally.

HMS FOWEY L 15, U 15

FOWEY commissioned with a Devonport crew for the Gulf and arrived 10.31 to exchange crews with FOLKESTONE. She served until 7.32, then re-commissioned at Colombo for further service. Her annual refit was also at Colombo 7-8.33, and she re-commissioned at Basra 11.34 to remain up the Gulf. Further refit at Colombo 5.36, and a new crew joined at Basra 17.11.36. In 4.38 FOWEY went to Malta to refit and re-arm, returning to the Gulf until 8.39 when she sailed for Bombay to repair.

Completing 9.9.39, FOWEY went to the Mediterranean until 11.39 when she went to Freetown, and returned to the UK with SL 11, to refit at Southampton.

FOWEY then joined Western Approaches Command, and took part in sinking U 55 30.1.40. She refitted at Devonport 5-6.40 returning to Liverpool on completion, and transferring to Rosyth Escort Force 11.40 to 6.41 before going back to the North Atlantic.

FOWEY in wartime, as a unit of 37 EG. This shows her in a somewhat odd guise, for she has the usual wartime improvements of two 4in HA, and four 20mm, 271 and 291 radars are fitted, also HF/DF on its tall mast aft. However, the sweepdeck bulwarks have been retained, although the DCTs are visible behind them. Oddest of all are the towing hoops, an unusual fitment in sloops and presumably provided due to the absence of rescue tugs from the Freetown convoy route which could require an escort to undertake a long tow of a damaged merchantman. FOWEY is under tow aft and also signals that she is at anchor, she is probably being swung for compass adjustment after refit.

Refits at Belfast 7.41, Liverpool 3-5.42 and 10.42 followed, and a major overhaul at Milford Haven 1-5.43. In 1.44 FOWEY went to the Mediterranean and had a repair period at Alexandria 2-3.44 before coming home in 4.44 for yet another refit at Milford Haven.

In 9.44 FOWEY became part of B 23 Group in Western Approaches and, at the end of the war, became static guardship at Stranraer, and later Larne, over surrendered U boats, a duty lasting until 12.45.

In 1.46 FOWEY transferred to Portsmouth and in 5.46 started refit prior to sale to Egypt, later cancelled. The ship was then sold in 10.46 to Wheelock, Marden & Co Ltd for £9,500 as *FOWLOCK* and is reported as being broken up at Mombasa in 1950.

HMS ROCHESTER L 50, U 50, F 50

ROCHESTER commissioned with a Chatham crew for service on the Africa Station where she remained until returning to re-commission 12.6.34 at Sheerness for further service, remaining based at Simonstown until 11.36. Even then, with the pattern of ships remaining on one Station for long periods, she re-commissioned 17.12.36 and returned South, refitting and re-arming at Simonstown 10.38-1.39.

When ROCHESTER re-commissioned at Sheerness 6.39, it was for the Gulf and she arrived at Aden 1.9.39, to return home 11.39 when she went to Freetown to collect convoy SL 12 and bring it to the UK and to refit herself on the Humber, prior to joining the 2nd Escort Vessel Division at Liverpool. She was in collision with *LONGFORD* 27.3.40 and subsequently repaired at Liverpool 6.40. ROCHESTER attacked and damaged U 94 7.5.41, refitted at Liverpool 6.41 and joined 37 EG, later transferring to 43 EG 9.41 for the Freetown route.

On 19.10.41 ROCHESTER took part in the sinking of U 204, and on 6.2.42 it was U 82 that succumbed. A further Tyne refit 5-7.42 did not break the run of success, for U 213 was the next victim on 31.7.42. Operation Torch followed and a transfer of the Group to Western Mediterranean Command until 2.43 when ROCHESTER returned to Liverpool and joined 39 EG. A Humber refit followed 5-6.43, and again ROCHESTER scored after refit, taking part in the sinking of U 135 15.7.43.

The original ROCHESTER, unfortunately with some interference from background, but clear enough to show her somewhat spartan original appearance.

ROCHESTER in 7.42, a superb picture. She is just out of refit with a new bridge, 20mm in its wings and 271 at the rear. 291 tops the masthead, a new HF/DF mast aft. The quadruple 0.5in mg remain in their bandstand amidships.

ROCHESTER immediately after conversion to a training ship, she still retains a 20mm on the focsle and two more on the quarterdeck. The new 277 radar adorns the bridge, and extra accommodation has been added in lieu of X gun.

Collision with HART 31.5.44 did not prevent her taking part in Operation Neptune in 6.44, then joining 41 EG based at Devonport. Thereafter ROCHESTER operated in the Channel Approaches and Biscay area to 10.44, a hunting ground of her predecessors. She went for refit on the Tyne 11.44-2.45, and was transferred to Portsmouth 3-7.45, emerging from refit as a dis-armed training ship for HMS DRYAD, the shore based Navigation School.

ROCHESTER in almost her final form with a new dark hull paintwork, outlined pennants (probably red outlined with white) and totally disarmed. Photograph taken 21.11.46 off Portsmouth.

ROCHESTER remained in this latest capacity, displaying ever changing Radar suites and paint schemes, until 9.49, going on the Sale List 7.9.49 after long and successful service. During her latter years she had become a major part of the Portsmouth scene, and her departure was regretted by many. She transferred to BISCO 1.51 and was broken up at Dunston on Tyne, arriving there 14.2.51.

SHOREHAM as completed. Note the complete lack of close range AA weapons and the old destroyer type 4in LA aft.

HMS SHOREHAM L 32, U 32

With a Portsmouth crew, SHOREHAM went "up the Gulf" where she was to serve both pre- and post-war. Initially the normal pattern was followed, refit at Colombo 10.32, an exchange of crews with TRIAD, the Senior Officer's ship, at Bombay and refit there 3-6.33. She docked at Bombay 11.33, re-commissioned 9.3.34 and had an annual refit, also at Bombay, 5-6.34. A further crew change at Bombay 19.3.36 and a refit 2.38 broke the Gulf service, before SHOREHAM went to Malta to refit and re-arm 1-4.39.

Even after this lengthy service, she returned to the Gulf, with Bombay refits 8.39 and 3-4.40. Wartime routine took her to the Red Sea, where amongst many activities she participated in the sinking of EVANGELISTA TORRICELLI on 23.6.40 and intercepted the EVROS 3.9.40. A Bombay refit in 10.40 was followed by return to the Red Sea and the interception of the German ODER on 23.3.41, fleeing the Italian colonies in the face of the British advance. SHOREHAM refitted again at Bombay 4-5.41 and then returned to the Gulf in time to take part in the occupation of Abadan, Operation Countenance. After this excitement, SHOREHAM went to Suez as AA guardship, but in 1.42 returned east to Colombo to join the Eastern Fleet for A/S work, one of the few escorts on Station.

A long refit at Bombay 10.42-2.43 preceded a move to Alexandria and the Levant Command until 9.43, when she reverted to the Eastern Fleet at Bombay, operating mainly on the west coast of India until 2.44 when she refitted at Capetown 3-6.44. On return, SHOREHAM shifted to the east coast of India to 8.44 followed by a brief spell at Bombay and more African service with a Durban refit 5-7.45. On completion SHOREHAM returned to Ceylon until 10.45, when she refitted at Bombay for a month preparing for peace time service. Surprise, surprise, orders after refit were for the Gulf again, where she operated until 7.46 when she came home to pay off for disposal.

Sold 4.11.46 for commercial service as JORGE F EL JOVEN, she does not appear to have operated commercially, for when she arrived at Zeebrugge for breaking in 11.50 she appeared unaltered from her dis-armed naval appearance other than for the name painted on her dilapidated hull.

courtesy L Van Ginderen

When SHOREHAM was sold post war in 11.46, she became JORGE F EL JOVEN. Here she proceeds under tow, forlorn and dirty, to the breakers in 11.50. From the lack of any visible alterations, she probably did not trade at all as she appears unchanged except for the mercantile name just visible forward.

Repeat SHOREHAM class

Name	Builder	Laid down	Launched	Completed
DUNDEE	Chatham/H Leslie	1.12.31	20.9.32	31.3.33
FALMOUTH	Devonport/H Leslie	31.8.31	19.4.32	27.10.32
MILFORD	Devonport/Yarrow	14.9.31	11.6.32	22.12.32
WESTON	Devonport/Yarrow	7.9.31	23.7.32	23.2.33

Displacement 1,060 tons standard. Dimensions length 281ft 4in, beam 35ft, draught 10ft 2in at full load.

Machinery two shaft, geared turbines, designed SHP 2,000 = 16.5 kts. Oil fuel 290 tons, consumption 0.6 tons per hour at 10 kts.

Initial armament one 4in HA, one 4in LA (not fitted in FALMOUTH) four 3pdr.

Changes during service

During refits in 1937/39 all four ships fitted a second 4in HA in lieu of the LA (in the case of FALMOUTH in lieu of the accommodation aft) and added a quadruple 0.5in mg.

DUNDEE, due to overseas service and early loss, had no known wartime alterations.

FALMOUTH shipped a second quadruple 0.5in mounting early in the war, possibly at Hong Kong 3.40 when ASDIC was also fitted, adding two single 20mm, increased to four later, at unknown dates.

MILFORD fitted a second quadruple 0.5in mounting, possibly at Simonstown 4.40, subsequently she shipped four single 20mm in lieu, later increased to five. Positive dates of change are unknown.

WESTON shipped a second quadruple 0.5in mounting probably at Portsmouth 9.39, then added two, later four, single 20mm.

Radar

All ships, other than DUNDEE, progressively fitted radar and improved the type, 286 then 271 being the normal pattern. Only WESTON is noted as fitting HF/DF.

A pre war DUNDEE showing the very basic outfit of the early sloop. The after LA gun clearly shows its old destroyer origin with its half shield, there is no close range armament whatever. The quarterdeck, with its solid bulwarks, houses the M/S winch, kites and otters of the wire sweep gear, with the heavy twin davits to handle them.

HMS DUNDEE L 84, U 84

DUNDEE completed and commissioned with a Portsmouth crew for service on the America & West Indies Station. During her service there, she attempted to salvage *BASIL* 11.6.33. She then returned to Portsmouth and recommissioned 12.10.35 for further service in the West Indies arriving at Bermuda 3.11.35 and returning to Portsmouth 14.3.38 to refit, re-arm and re-commission 5.7.38 for her old Station.

She grounded at Ellis Bay 20.9.38, remaining fast for 4 days. On the outbreak of war, she was Trinidad-based for patrol duties to 7.40 thence to Bermuda for refit to 27.8.40. On completion, DUNDEE sailed for Sydney CB to escort convoy SC 3 from that port, sailing 2.9.40. DUNDEE was torpedoed by U 48 14.9.40 while still with the convoy, and sank.

HMS FALMOUTH L 34, U 34, F 34

FALMOUTH commissioned with a Devonport crew for service in China as Admiral's Despatch Vessel. In this role she omitted the armament and M/S gear aft and was completed with additional accommodation for the Commander-in-Chief, his wife, retinue and staff and was used for official visits to ports too small to permit entry of the cruiser flagship. She remained in this role until the outbreak of war.

Re-commissioned at Singapore with a fresh crew 18.2.35, she took part in the evacuation of Britons from Shanghai 8.37, shortly before re-commissioning again at Hong Kong 8.10.37.

Refitted and re-armed at Hong Kong, including the fitting of ASDIC in 3.40, FALMOUTH was thereafter Singapore-based for patrol work until going to Colombo 5.40. She then went to the Persian Gulf where she remained until sent to Bombay for refit 12.40, sinking the Italian submarine GALVANI 23.6.40.

Returning to the Gulf, FALMOUTH took part in the occupation of Basra 28.4.41 and of Abadan (Operation Countenance) 25.8.41, after which she towed *BRONTE* to Karachi, departing from Basra on 4.9.41.

Sent to Suez 10.41, FALMOUTH returned to Ceylon on the outbreak of war in the East Indies and remained principally based there until 7.42 when she refitted at Bombay to 11.42, thereafter taking up duty with the Gulf and Aden convoys to and from Bombay.

A very clear 1943 view of FALMOUTH, taken at Bombay. Just visible beyond her focsle is a mast and funnel and a bow with pennant 85, which identifies that ship as the corvette VERBENA and the probable date of the photograph is 1.43. FALMOUTH has two 4in HA, single 20mm in the bridge wings and quadruple 0.5in in the bandstands. 291 and 271 radars are fitted.

FALMOUTH refitted at Durban 6.43, then began duty with east coast Africa convoys for the rest of the war, refitting at Simonstown 12.43-2.44 and 3-6.45 after which she returned to Colombo and for operations off Rangoon and the Burmese coast. In early 1946 she again operated in the Persian Gulf, and finally returned at the end of that year to the UK for the first time since completion, to pay off to reserve and await disposal.

FALMOUTH was reprieved from the breaker's yard by the decision to gut and refit her as a new drillship for the Tyne Division RNVR, in which guise she was re-named CALLIOPE and stationed at Newcastle in 1.52, where she remained until towed from her berth on 30.4.68 to be broken up at Blyth.

HMS MILFORD L 51, U 51

Commissioning with a Portsmouth crew for the Africa Station, MILFORD remained there until returning to Portsmouth to refit and re-commission on 11.5.35 for further service on the Coast. She returned to Portsmouth to refit and re-arm in 10.37 for further Africa service, but this time with a Devonport crew who took over 29.12.37. Further re-arming took place at Simonstown 1 and 5.39 with the ship remaining on the Station, which was re-titled the South Atlantic Command in 9.39.

MILFORD was Freetown-based from the outbreak of war until 1.40 when she went south to refit at Simonstown, returning to Freetown in time for operations against Dakar in 7.40 and 9.40 (Operation Menace). Taking part in the landings at Libreville 11.40, she sank the French submarine PONCELET 8.11.40.

MILFORD remained Freetown-based to 5.41 when she refitted at Simonstown, thereafter being based there to 12.41, when she returned to Freetown until 7.42. In 7.42 she had a welcome and unusual refit, going to the Brazilian Naval Yard at Rio de Janeiro until 9.42 before returning to Freetown.

The wartime MILFORD, little changed from pre-war except for the 4in HA aft, and loss of her M/S gear and mainmast.

MILFORD returned to the UK for refit on the Clyde 9.43 and to join 40 EG. Unfortunately the old hull was not up to further operational service and she was laid up at Ardrossan with defects 12.43, being taken in hand on Teesside 6.44 to refit as a submarine target ship and escort.

Attached to the 10th Submarine Flotilla at Rothesay, MILFORD continued that duty post war, refitting on the Clyde in 12.45 and then transferred to the 7th Submarine Flotilla at Portsmouth until retiring to reserve. Transferred to BISCO 3.6.49, MILFORD was broken up at Hayle by T W Ward Ltd.

MILFORD in 1944, now serving as a target and escort for submarines. As can be seen she has updated her close range armament by fitting four single 20mm, and 271 radar is fitted. The fixed bulwarks of the sweep deck have been cut right back almost to the break of the focsle, and two heavy davits fitted at the after end of the focsle deck to recover practice torpedoes. The A/S armament has been reduced to two DCT and the rails aft.

WESTON in pre war configuration, apparently coming home to pay off.   Wright & Logan

HMS WESTON L 72, U 72

Originally intended to be named WESTON-SUPER-MARE, the ship rejoiced throughout her career in the nickname "Aggie on horseback", a reference to the late Agatha Weston who provided temperance social facilities for Victorian seamen.

WESTON commissioned with a Portsmouth crew for the Africa Station, serving there until re-commissioned at Gibraltar 31.8.35 for a further two years' service in the Red Sea. She re-commissioned at Malta 18.8.37 for further Red Sea service, then returned to Malta for long refit 2-6.39, transferring to Portsmouth to complete the refit, achieved during 9.39.

Allocated to the Rosyth Escort Force, WESTON commenced her service by running aground 13.9.39 before starting her duty with East Coast and, later, OA convoys. During this service, she sank U 13 on 31.5.40. Attached to the Northern Escort Force 1.41 and then Londonderry Sloop Division 7.41 and 42 EG 1.42, WESTON was refitted at Dundee 6-9.42 and Belfast 6-8.43 after which she returned to Freetown as part of the West Africa Command.

Serving at Freetown, WESTON joined 55 EG 3.44, and had a welcome break from the Coast 10-12.44 when she refitted at Bermuda before returning to Portsmouth 1.45 for further work and home leave. Allocated again to West Africa, she remained however at Portsmouth until passing to reserve 6.45. Allocated to BISCO 22.5.47, WESTON went to Gelleswick Bay for breaking.

The wartime WESTON. Note the new bridge with 271 radar and 20mm. Hedgehog to port of the forward 4in HA, a further 4in HA aft, and a tall HF/DF mast.

GRIMSBY class

Name	Builder	Laid down	Launched	Completed
ABERDEEN	Devonport/Thornycroft	12.6.35	22.1.36	17.9.36
DEPTFORD	Chatham/White	30.4.34	5.2.35	20.8.35
FLEETWOOD	Devonport/Thornycroft	14.8.35	24.3.36	19.11.36
GRIMSBY	Devonport/White	23.1.33	19.7.33	17.5.34
LEITH	Devonport/White	6.2.33	9.9.33	12.7.34
LONDONDERRY	Devonport/White	11.6.34	16.1.35	20.9.35
LOWESTOFT	Devonport/White	21.8.33	11.4.34	22.11.34
WELLINGTON	Devonport/White	25.9.33	29.5.34	24.1.35

Displacement 990 tons standard. Dimensions length 266ft 3in, beam 36ft, draught 9ft 6in at full load.

Machinery two shaft, geared turbines, designed SHP 2,000 = 16.5 kts. Oil fuel 300 tons (337 in LEITH and WELLINGTON), consumption 0.5 tons per hour at 10 kts.

Initial armament two 4.7in LA, one 3in HA (two 4in HA in ABERDEEN, two twin 4in HA in FLEETWOOD). Four 3pdr saluting guns in all except ABERDEEN, one quadruple 0.5in mg in ABERDEEN and FLEETWOOD.

Changes during service

ABERDEEN added two, later four, single 20mm at unknown dates, and fitted Hedgehog in 4.42.

DEPTFORD retained her 4.7in throughout, adding two quadruple 0.5in mg in her pre war re-arming. In 1.42 added two single 20mm, increased to 4 in 10.42. In 1943 the quadruple 0.5in were landed and two further single 20mm shipped. Hedgehog was fitted in mid 1942.

FLEETWOOD retained her twin 4in, adding a second quadruple 0.5in in 1938. In 1941 three single 20mm were added, while a further three were shipped in 1942 in lieu of the 0.5in mountings. Hedgehog was fitted 6.43.

GRIMSBY Due to her early loss it is probable that no changes were made to her original armament.

LEITH retained her 4.7in, landing her quadruple 0.5in in 10.42 in exchange for four single 20mm, later increased to six. Hedgehog was fitted 6.43.

LONDONDERRY re-armed with two twin 4in HA in lieu of her 4.7in and 3in, added two single 20mm in lieu of her quadruple 0.5in, and increased the number of 20mm to six by mid 1943. Hedgehog was fitted 10.43.

LOWESTOFT re-armed with two twin 4in HA in late 1939, also fitting a quadruple 0.5in mg. A second quadruple 0.5in and two single 20mm were added in 1941, the 0.5in being landed and four further 20mm fitted during 1942. Hedgehog was fitted 8.43.

WELLINGTON had few changes, two single 20mm being added, later increased to six. The 3in HA was landed and Hedgehog fitted in mid 1943.

Radar.

ABERDEEN fitted HF/DF and 271 in 4.42

DEPTFORD fitted 286 early in the war, replaced by 271 in 3.42. HF/DF was fitted in 3.42.

FLEETWOOD fitted 286M in early 1941, replaced with 286P. HF/DF was added 7.42, and 271 10.42.

GRIMSBY was probably not fitted with radar due to her early loss.

LEITH acquired 286 early in 1941, replaced by 291 at an unknown date, 271 and HF/DF added 4.42.

LONDONDERRY had her early war 286 replaced by 291 in 1943, and had fitted 271 in 8.42. HF/DF added mid 1942.

LOWESTOFT probably had 286, HF/DF, 271 and 291 were fitted early in 1943.

WELLINGTON 271 and HF/DF were fitted 6.42.

ABERDEEN in all her glory as a Commander-in-Chief's yacht, with white hull and buff funnel. Both the forward guns are unshielded, the quadruple 0.5in mg is very evident amidships. The photograph was taken when the ship attended the Coronation Review of 1937.

HMS ABERDEEN L 97, U 97

ABERDEEN was selected to complete as a Despatch Vessel to serve as alternate flagship on the Mediterranean Station, and was accordingly fitted with extra accommodation aft in lieu of her M/S gear and after armament. The ship finally commissioned after trials 15.9.36 at Devonport, and visited Aberdeen after her Portland work up before proceeding to Malta where she arrived for service 5.11.

ABERDEEN cruised extensively in the Mediterranean until 5.37, when she returned to Portsmouth for the Coronation Review and to give brief leave at Devonport afterwards. Thereafter she returned to station, and to her intensive cruising programme with the Flag, which continued until a six week refit at Malta 2.38. A further month's refit at Malta in late 8.38 was interrupted by the Munich crisis, and ABERDEEN proceeded rapidly to Suez where she became guardship during the crisis, to return to Malta and her normal routine in mid 10.38. During an official visit to St Tropez 3.39, the ship dragged and went ashore, fortunately on a mud bottom, sustaining slight damage for which she was docked at Malta in the latter part of 4.39. Her re-armament, planned for this period, was deferred due to the international situation and she retained her single 4in HA, although acquiring a third such mounting in X position.

On the outbreak of war, ABERDEEN returned to the UK and initially operated in the Channel area until the French surrender closed that approach when, still part of 1st Escort Vessel Division of Western Approaches Command, she transferred to the northern route operating out of Rosyth until 11.40 when she was moved to the Liverpool Sloop Division. Operating out of Liverpool, ABERDEEN suffered minor damage in a collision with COLUMBINE 22.12.40, but did not enter dockyard hands until 3.41 when she had a month's refit. On completion, ABERDEEN escorted OG 57 to Gibraltar and then took passage to Halifax NS to bring home an HX convoy.

Briefly attached to the new Newfoundland Command in 6.41, ABERDEEN then transferred to 41 EG of Western Approaches for long distance escort duty on the Freetown route, which continued until 11.42 when she formed part of the escort for the assault convoy, KMF 1, for Operation Torch. On completion of the assault, ABERDEEN brought back the first empty troop convoy, MKF 1Y, and went to refit on the Tyne from 4.12.42-23.2.43, being administered by 45 EG during this period.

After work up following her refit, ABERDEEN briefly joined 40 EG and went to St Johns NF to collect her first convoy, HX 229A. During the passage of this convoy it was diverted so far north to avoid attack that it became embroiled in the ice edge, losing the whale factory ship *SVEND FOYN* to ice damage. ABERDEEN herself went aground on an ice ledge during the night of 19.3, and was indeed fortunate to get herself off by working her engines astern, although with serious bottom damage and the loss of her ASDIC dome. The damage was repaired at Liverpool 26.3-30.5, followed by work up at Tobermory.

ABERDEEN in 1943, much changed from the previous photograph. Probably taken at Freetown. Note the replacement of A gun with Hedgehog, the fitting of 271 radar above the bridge and 291 at the masthead, and the mainmast rigged well aft with the Adcock aerial of HF/DF at the top. W/T aerials are now rigged from the spreader aft which replaces the mainmast, just forward of this can be seen the port bandstand with a 20mm mounted thereon.

The pennant is a mystery, either she has retained her pre 4.40 L pennant (unlikely) or the censor has converted the U into an L. Quite positively, the photograph is dated after 4.42.

ABERDEEN returned to the West African coast, joining the new West African Command and, until 4.44, escorted convoys south from Freetown to Nigerian ports, or the local end of the OS/SL series. In 4.44, ABERDEEN returned home with convoy SL 154 and went for refit to Devonport for three weeks while granting home leave. Her ship's company must have been pleased on return from leave when, still requiring further work, ABERDEEN was sent to Bermuda for completion, an interlude that lasted until the end of 8.44.

Returning to the West Coast, ABERDEEN operated thereafter from Freetown until 8.45 when, wartime escorting and the immediate post war Air/Sea Rescue patrols over, the old ship steamed to Gibraltar where she was laid up in reserve, remaining there until the end of 1946 when she was brought home to Devonport and placed on the disposal list. Allocated to BISCO at the end of 1948, ABERDEEN arrived at Hayle on 19.1.49 to be broken up.

DEPTFORD painted ready for overseas service with white hull and buff superstructure. The two calibre armament, 4.7in LA in A and X and 3in HA in B is very clearly shown. Note that she is fully equipped with M/S gear, but no trace at all of A/S armament.

HMS DEPTFORD L 53, U 53

Built at Chatham, DEPTFORD commissioned there with a Chatham crew 14.8.35 and visited her London Borough namesake for two days prior to sailing for Muscat and the Persian Gulf. Arriving at Muscat 14.11, crews exchanged ships between LUPIN and DEPTFORD, which must have been a sore disappointment for the Chatham men leaving a brand new ship for a 1916 veteran. Thereafter, DEPTFORD cruised in the Gulf until 3.36 when she went to Bombay for her bi-annual docking. This lasted only 14 days, the more lengthy annual refit taking place, again at Bombay, 19.9-10.11.36.

DEPTFORD returned to Bombay 4.4.37, not only to dock but also to re-commission with a new crew sent out in the transport NEURALIA, this being done 12.4.37. The new crew took the ship back to the Gulf until late 11.37, when a six week refit commenced in commercial hands at Colombo, after which she returned to station.

9.38 saw a month's refit at Bombay, during which the ship was fitted with ASDIC for the first time; it was intended that a refit at Malta should see the ship re-armed, probably with twin 4in HA, in 5.39. In the event, after further Gulf service, DEPTFORD did go to Malta but only for a three weeks docking and minor refit; the international position did not permit the planned three and a half months' work.

On completion, DEPTFORD served briefly in the Gulf and then sailed for Colombo and Singapore, arriving at the latter 25.8.39 only to be ordered back due to the imminence of war. Pausing only briefly during the passage, DEPTFORD returned to the UK for the first time since completion, arriving at Portsmouth 16.11 where she fitted her depth charge equipment and generally completed for war to join 2 EV Division of Western Approaches based on Liverpool.

Based on Liverpool, DEPTFORD escorted oceanic convoys in the UK Approaches, colliding with ANTIGUA whilst with OB 84 on 3.2.40; the damage was minor as no dockyard time seems to have resulted. In mid 8.40, DEPTFORD transferred briefly to Rosyth, but very soon returned to Liverpool as part of the Sloop Division based there to handle North Atlantic traffic. In 2.41 she refitted at Liverpool and then transferred to the Gibraltar route until the end of the year, celebrating the approach of Christmas by participating in the sinking of U 567 on 21.12 and colliding with STORK the next day. Unfortunately, the subsequent repairs awaited return to Liverpool on 28.12, so that Christmas was a little late!

Completing repair work in mid 3.42, DEPTFORD joined 36 EG based at Liverpool for the Freetown convoy route until 11.42 when she escorted the stores convoy for the invasion of North Africa, Operation Torch, thereafter operating along the Algerian coast until 9.12 when she went aground at Algiers. Ashore until 14.12, DEPTFORD returned to Liverpool in convoy MKS 8 to repair damage and refit, a process that took until late 9.43 followed by the inevitable Tobermory work up.

After work up, DEPTFORD joined 37 EG in the Mediterranean Fleet and took her first convoy to Gibraltar, arriving there 7.11 to commence escort duty in the Mediterranean until the end of 12.44, being based at Alexandria and, latterly, Taranto. Returning to Portsmouth in time for Christmas 1944, DEPTFORD refitted and joined Portsmouth Command for Channel convoy escort duty until the end of the war. She sailed to Cardiff, arriving 8.6.45 to de-store, and then went into reserve at Milford Haven 3.7, where she was to lie for over two years before being allocated to BISCO 8.3.48 to be broken up locally.

HMS FLEETWOOD L 47, U 47, F 47

FLEETWOOD was chosen as a testbed for new armament, the twin 4in HA mounting later to become standard for all sloop units. Completed at Devonport, she was manned from Portsmouth and commissioned 17.11.36 for extensive trials. Her first engagement after work up was not the now customary visit to her name port, but an extended winter cruise to Gibraltar, Tangier and Cartagena 2-3.37. On return, and following a docking period at Portsmouth, FLEETWOOD visited London, Fleetwood, and the Scilly Isles topped off with a further period at Portsmouth and Invergordon for more experimental work.

Refitted at Portsmouth 15.11.37-7.2.38, FLEETWOOD re-commissioned, leaving the Home Fleet to which she had been attached and completing for service in the Red Sea, arriving at Aden 27.3 and serving in that area until refit at Malta where she arrived 17.10. The Malta interlude was brief, FLEETWOOD sailed for the Red Sea 7.12 and remained there until recalled for a hurried docking at Malta prior to returning to Aden in 4.39. When war commenced, FLEETWOOD was ordered to Port Said but paused in the Mediterranean only briefly, arriving at Portland 11.10.

With her modern HA armament, FLEETWOOD was assigned to duty on the East coast for the defence of the Methil/Southend convoy route, refitting at Dundee 10.11-21.12.39. In 4.40, she took part in the Norwegian campaign, remaining there until the final evacuation from Harstad 1.6, then returning to Rosyth for the remainder of 1940.

At the end of 1.41, as part of the Northern Escort Force, FLEETWOOD was re-allocated to Londonderry and took up duty with the North Atlantic convoys until 4.41 when she joined the 1st AA Division of Western Approaches Command. Still based on Londonderry and Liverpool, FLEETWOOD now concentrated on the defence of convoys against Luftwaffe attack in the approaches to the UK, a duty that continued until 6.41 when the ship joined the newly formed Newfoundland Command. Service here was brief, a Middlesbrough refit 8-9.41 leading to her re-joining Western Approaches in the Sloop Division and duty with the Freetown convoy route, changing organisation to 44 EG in 11.41. This was interrupted by repairs at Liverpool for almost three months in 6-9.42 following collision with the tanker *OILRELIANCE* 2.6.42 while with convoy OS 30.

FLEETWOOD in war time form. Note the twin 4in HA now mounted fore and aft, B position carries close range weapons. There is HF/DF aft, 271 on the bridge and 291 at the masthead.

Following repair, FLEETWOOD resumed the Freetown route briefly, then became involved in the follow up to Operation Torch, escorting back-up convoys to and from North Africa, being attached to the Western Mediterranean Fleet as part of 43 EG for the first quarter of 1943. In 4.43 FLEETWOOD transferred to 39 EG and returned to the Londonderry base, having her first success of the war on 11.5 when she sank U 528 while escorting convoy OS 47.

A refit at Dundee, and a change of crew, took place in 7-8.43 and, after work up at Tobermory, FLEETWOOD escorted one convoy out to Gibraltar and was then employed on A/S patrol from that base, sinking U 340 on 1.11. Unfortunately, while at Gibraltar she was firstly in collision with the US patrol craft PC 473 and then struck a jetty while berthing on 17.11, damage that confined her to Gibraltar Dockyard until the end of 3.44.

During her repairs, FLEETWOOD came under 41 EG, Mediterranean Fleet, and she remained with that Group after completion, escorting convoys through the Mediterranean to Port Said.

After a refit at Haifa in 9.44, FLEETWOOD transferred to 50 EG briefly, then came home to join Portsmouth Command for Channel duty at the end of 1944. Post 5.45 the ship remained in Portsmouth Command until 4.8.45 when she arrived on Teesside to lay up in reserve, and probable disposal.

Wright & Logan

FLEETWOOD post war as a radar training ship. Her radar fit varied so frequently that it is not practical to illustrate all her guises, but this view is typical of her in the post war era operating from Portsmouth as a tender for trials and training purposes.

A change of plans, and role, gave FLEETWOOD new life; she was re-activated and taken to Portsmouth 2.46 to disarm and refit and become the Radar training ship for the Royal Navy. In this duty, with an ever changing radar suite, FLEETWOOD became an integral part of the Portsmouth scene, operating actively until her final disposal in 10.59, the old ship arriving at Gateshead on Tyne 10.10.59 to be broken up.

HMS GRIMSBY L 16, U 16

GRIMSBY commissioned at Devonport with a Portsmouth crew on completion of her acceptance trials, despite dragging her anchor during a squall and suffering slight structural damage in collision with the tug *FAIRPLAY*. After visiting both Grimsby and Cleethorpes GRIMSBY sailed from Devonport escorting BARNET bound for Hong Kong, where she arrived on 5.11.34. On arrival she was taken in hand for docking, and also to fit her minesweeping gear, drawn from stock in the Dockyard. Thereafter she was employed at Hong Kong and along the Chinese coast, with her annual dockings at Hong Kong.

GRIMSBY went to Singapore in 9.36, and re-commissioned there 8.10 with a new crew sent out from the Portsmouth depot, then returned to Hong Kong to resume her usual duties, which included frequent anti-piracy patrols.

In addition to her usual annual refits at Hong Kong, slightly unusual work was done in 12.37 when the Hong Kong Dockyard spent three weeks fitting the ship for "service in cold climates", presumably with a possible over-wintering at Wei-Hai-Wei in mind; GRIMSBY was not to return to the UK until 10.39 and does not seem to have gone north of Rosyth even then! In 2.39 GRIMSBY went to Singapore, both for her annual docking and for re-arming and general updating in common with all the earlier sloops. She remained at Singapore 28.2-26.7, re-commissioning on 16.3. On completion of the refit, the ship transferred to the East Indies Station, arriving at Colombo 2.8. After a short period in the Persian Gulf,

GRIMSBY shortly after completion, carrying out mine laying trials. She has landed her X gun to cut top weight, and the mine rails have been laid on the quarterdeck.

GRIMSBY on completion, showing her fitted out for minelaying with the mining sponsons on the quarter. She has not shipped her mine rails, nor has she landed the needed top weight to compensate for a load of mines. Compare this photograph with the preceding one, which shows her in the minelayer role. The heavy davits abeam X gun are for loading mines. Note the 3in HA in B position, and the destroyer type 4.7in LA mountings in A and X.

service terminated by the outbreak of war, GRIMSBY returned to Portsmouth to serve with the Rosyth Escort Force. The outbreak of war, and the need for the ship to serve at once on the East coast, prevented the proposed re-arming with two twin 4in HA in lieu of her single 4.7in LA.

Service at Rosyth included a month's refit at Leith in 4.40, and terminated when GRIMSBY was ordered to join the Mediterranean Fleet, leaving Devonport on 20.5 and arriving at Port Said to join the Red Sea Force on 31.5. GRIMSBY then became heavily involved in escorting convoys between Suez and Aden, duty that occupied her until early 3.41 when she arrived at Port Said to rejoin the main Fleet. GRIMSBY took part in convoying ships to and from Crete in Operation Demon, and was then lost to air attack off Tobruk on 25.5.41.

A final GRIMSBY photograph; she has now dispensed with the mine laying role, having removed both the loading davits and the sponsons in the hull and re-stowed her M/S gear. The A/S armament, the most important item, is still lacking!

HMS LEITH L 36, U 36

LEITH completed at Devonport and, after trials with a local crew, commissioned there 10.7.34 with a Chatham crew for service on the New Zealand Station, for which she sailed on 13.8 after a courtesy visit to the port of Leith, arriving on station at Auckland 13.11. LEITH's duty in New Zealand waters comprised visits to the principal ports, and also regular courtesy calls and cruises to Fiji, Samoa and adjacent island groups, the ship being based at Auckland.

A good pre-war view of LEITH preparing for foreign service. Note the 3 pdr saluting guns on the shelter deck abaft the break of the focsle, a similar pair are mounted to port.

A new crew for LEITH was sent out to Wellington, and the ship re-commissioned there 17.12.36 after which she conveyed the Commissioner for the Gilbert & Ellice Islands to the Phoenix Islands where due proclamation of British sovereignty was made during the second week of February 1937. Quite why this was necessary at so late a date is unknown. The following six months were taken up with extended cruises amongst the South West Pacific islands, the ship not returning to Auckland until 18.9.

Further cruising in 1-2.38 was interrupted by the need to go to Sydney NSW for work on defective shaft bearings, presumably beyond the resources of the Auckland Dockyard, this being done during a six week visit 5-6.38. On completion, further island cruising, including carrying the Queen of Tonga on Royal visits to outlying islands, preceded LEITH returning to Auckland at the end of 9.38.

LEITH re-commissioned at Wellington with a further Chatham crew 27.7.39 and, on the outbreak of war, proceeded to Jervis Bay and thence to Singapore where she came under the orders of Commander-in-Chief, China Station, operating out of Penang until 7.11. From Penang LEITH sailed for home, being diverted from Gibraltar to Freetown to pick up the homeward convoy SL 14, leaving Freetown 26.12 and arriving at Penarth for refit 13.1.40.

On completion of refit 2.2.40, LEITH proceeded to Liverpool to join Western Approaches Command, principally handling the UK/Gibraltar convoy traffic involving 4 to 5 days out from Liverpool before exchanging to an inward bound convoy. This pattern continued until mid 7.40 when LEITH transferred to Rosyth only to burn out No 1 boiler within days, involving a month's repair at Belfast until 12.8 after which the ship returned to her Liverpool base and became involved in North Atlantic convoys in addition to the Gibraltar traffic, making a through passage to Sydney NS in 11.40.

During 4.41 LEITH escorted Irish Sea convoys, and then undertook a lengthy refit at Avonmouth from 17.4-6.6.41, after which the ship joined the Newfoundland Escort Force which had just been formed. Service here was brief, two convoys only, for LEITH then returned to the Gibraltar passage in 8.41, in which operational area she was to remain for much of the rest of the war.

A wartime view of LEITH, her pennant number is unusually far forward for British service, presumably the result of a dockyard new to naval practice. Note the HF/DF mast aft, and the modern radar fit, although the 0.5in mg mountings appear to have been retained.

With the re-organisation of Western Approaches Command, LEITH joined 38 Escort Group based on Londonderry, transferring to 41 Escort Group in 8.43, prior to which she had taken part in Operation Torch, escorting the follow-up stores convoy KMS 2 and then operating along the North African coast in the early months of 1943.

Service with 41 EG transferred LEITH to the Freetown route, operating with the OS and SL convoy series until 8.44 when 41 EG transferred from Western Approaches to the Mediterranean Fleet to be based at Gibraltar. LEITH was, in fact, feeling the effect of strenuous service and in 9.44 commenced an extended refit at Gibraltar that lasted until 1.45, the ship then returning to Portsmouth to rejoin 38 EG, by now based on that port. LEITH operated thereafter in the Channel as a local escort for the large convoys now again using that passage to the UK, involving her in continuous running until 6.45 when she paid off to reserve at Rosyth.

LEITH in her final guise as the R Danish Navy GALATHEA, having been re-purchased after sale for commercial service by Britain.

Put up for sale as part of the post-war disposals, LEITH was sold for commercial service as *BYRON* 25.11.46, later being re-named *FRIENDSHIP*. Normally this would have been the end of her naval association. However, in an unusual move, the Royal Danish Navy purchased the ship in 1949 and re-commissioned her as the survey ship GALATHEA, in which guise she served on until sold for breaking up in 1955.

LONDONDERRY as built and prepared for foreign service, note that no close range AA armament is fitted, neither is there any A/S equipment visible.

HMS LONDONDERRY L 76, U 76

LONDONDERRY commissioned with a West Country crew 17.9.35 and, without the formality of a visit to her namesake, sailed for duty with the Red Sea Division 30.9 arriving on station in time to become guardship at Suez for ten weeks during the Abyssinian crisis. Thereafter, she remained in the Red Sea, except for her periodic dockings and storing at Malta, until 3.38 re-commissioning at Malta on 20.9.37 with a new crew sent out from Devonport.

In 4.38, LONDONDERRY was ordered to Simonstown, then under British control, for her annual refit, arriving there 3.5 and remaining until 21.6. She then joined the South Atlantic Station and cruised north to Freetown and Dakar, returning to Simonstown 6.11 to remain there until 5.1.39. Local cruising occupied her until 13.2.39 when she was taken in hand in the dockyard at Simonstown for re-arming with the twin 4in now becoming standard for the modern sloops, a refit that lasted until 18.7.39.

LONDONDERRY remained at Simonstown, patrolling after the outbreak of war until 11.10 when she went north to Freetown where she was based as a local escort and minesweeper until the arrival of trawlers for that duty, finally sailing as escort to SL 10 on 24.11 and arriving at Devonport for a brief refit 12.12 on completion of which she joined the Rosyth Escort Force.

LONDONDERRY remained based at Rosyth, escorting East coast convoys until the end of 1940 when she was transferred to Western Approaches 1st AA Division until the end of 5.41. During 6.41 LONDONDERRY briefly served with Newfoundland Command, returning to Londonderry in 7.41 to join the Sloop Division based there to cover the inward end of the Freetown and Gibraltar convoy routes.

In 12.41 LONDONDERRY joined 40 EG for the Freetown route, refitting at Avonmouth 4-6.42. She was diverted from her normal duties 10.42 to 1.43 to escort North African convoys in the post Operation Torch period, including the search for CORNCRAKE when that ship foundered in the heavy weather of 1.43.

Sailed from Londonderry 30.1.43 to meet and escort HX 224, LONDONDERRY was heavily damaged (her stern blown off by prematurely detonated depth charges) on 4.2.43, being towed to Londonderry for temporary repairs that lasted until 12.3. Permanent repair was to be done at Devonport and LONDONDERRY was towed there by the tug HUDSON arriving 14.3 and remaining under repair until 17.11.43.

Following a Tobermory work up, LONDONDERRY then joined 39 EG based at Londonderry and escorted trade convoys to and from Gibraltar until the end of 4.44. On 5.2 she collided with the trawler CAPE ARGONA, minor damage only, and in 5.44 she became part of 41 EG based at Devonport and under that Command. LONDONDERRY remained part of that Group until 5.45 escorting the rapidly increasing traffic using the Channel approach after the clearance of northern France.

LONDONDERRY paid off shortly after the end of the European war, arriving at Milford Haven to lay up 22.5.45, remaining there until allocated to BISCO 8.3.48, and arriving at Llanelly 8.6.48 to be scrapped.

A re-armed LOWESTOFT now equipped with twin 4in HA fore and aft, she has also shipped single 20mm in the bridge wings and a single 2 pdr right aft on the quarterdeck. The 286 mattress aerial is very clearly shown, the only radar then fitted. Note the unusual towing hoops fitted aft, a feature rarely seen in sloops.

HMS LOWESTOFT L 59, U 59

Built at Devonport and manned locally for trials, LOWESTOFT commissioned with a Devonport crew 20.11.34 for service on the China Station, proceeding there 13.12 after a courtesy call on the port of Lowestoft, arriving at Hong Kong 15.2.35 in dire need of repairs to defective boilers, work which lasted until 17.5. The usual round of China coast patrols and visits was varied for LOWESTOFT by calls in Borneo, Sarawak and Hainan in 1.36 and a Japanese visit in 8.36. The first commission ended when the ship went to Singapore to re-commission with a new crew sent out from Devonport, the transfer taking place 4.4.37.

The usual annual dockings at Hong Kong were varied by a longer refit at Singapore 9-11.38 followed by a visit to the Philippines and further Hong Kong based service. In 7.39 LOWESTOFT was taken in hand at Hong Kong for re-arming, shipping two twin 4in HA in lieu of her 4.7in LA, work that completed 8.12.39 when she sailed for home and war service. On passage LOWESTOFT collected her first convoy, HG 17F, at Gibraltar 29.1.40 and arrived at Devonport for a brief docking 5.2 before joining the Rosyth Escort Force.

Arriving at Rosyth 3.3.40, LOWESTOFT commenced escorting convoys north from the Forth across the Moray Firth, a hunting ground for long range aircraft even then, this work continuing until 10.40 when the ship moved to East coast work. On 5.1.41 this was terminated when LOWESTOFT was mined, involving repair at Chatham until 3.10.41 and then return to service at Rosyth until 4.42.

In 4.42 LOWESTOFT sailed to Londonderry to join 45 EG in Western Approaches Command escorting Freetown convoys until a collision with the Free French LEOPARD 12.7.42 which put her under repair at Gibraltar until late 11.42. Even then the work was incomplete, Gibraltar being short on staff and materials and very much involved with Operation Torch in late 1942, so that LOWESTOFT took passage in convoy MKS 2A to Falmouth and further repair until 4.43.

On completion of work up, LOWESTOFT joined 42 EG based at Londonderry and for three months escorted Gibraltar bound convoys, then transferred to West Africa Command based on Freetown. Duty here involved West Coast convoys south from Freetown, arduous work given the very limited refit facilities available at that port. LOWESTOFT remained on the West Coast until 6.44, by which time a full refit must have been badly needed, and the ship came home with convoy SL 161 to refit at the wartime refit base of Dunstaffnage, almost as remote but somewhat cooler than Freetown!

On completion, and with a new crew worked up at Tobermory, LOWESTOFT returned to the Coast, in her old 57 EG, to operate from Freetown in 12.44 and 1.45 before the whole Group transferred to Gibraltar to handle that end of the Gibraltar/UK route.

LOWESTOFT served out the remainder of the war at Gibraltar, returning to Devonport 6.45 to de-store prior to going into reserve at Milford Haven. From Milford Haven, LOWESTOFT was sold 4.10.46 for commercial service as *MIRAFLORES,* and traded as such for a number of years, finally arriving at Zeebrugge 5.8.55 for scrapping by Belgian shipbreakers.

WELLINGTON pre war as completed.   WSPL

HMS WELLINGTON L 65, U 65

WELLINGTON was the first of the post 1926 sloops not to conform to the custom of using British seaport names, due to the intent that she should serve on the New Zealand Station on completion. Built at Devonport, she commissioned there 22.1.35 and, after full trials, sailed for New Zealand 5.2, arriving on station at Wellington 13.5.

WELLINGTON, after a two month stay at Auckland, then embarked on a lengthy cruise covering the South West Pacific islands, returning to Auckland in late 9.35, a pattern to be repeated in subsequent years.

WELLINGTON re-commissioned at Wellington 26.8.37 for further service in New Zealand. In the same manner as LOWESTOFT, shaft problems were discovered in early 1938 and the ship went to Sydney NSW for a six week docking in 5.38 for repairs. Presumably the Auckland Dockyard was overloaded with work for a similar docking period at Sydney took place in 2.39 prior to the mid 1939 islands cruise which lasted until 19.8 when WELLINGTON was ordered to Auckland to prepare for war.

WELLINGTON sailed from Auckland 3.9 for her war station at Singapore, via Townsville, arriving 19.9 to serve based at Penang until 2.11 when ordered to the UK. Passing Gibraltar 9.12, WELLINGTON was sent to Freetown to collect convoy SL 13 for the UK, finally arriving at Cardiff 9.1.40 for a brief refit, on completion of which she joined Western Approaches Command in the 1st Escort Vessel Division based at Devonport to cover Channel and Gibraltar convoys.

With the suspension of major convoy traffic in the Channel after the surrender of France, WELLINGTON, in common with other escorts, moved her base to Liverpool and operated from there with North Atlantic convoys as well as the Gibraltar traffic. While escorting convoy HG 43 home from Gibraltar, she was in collision with *SARASTONE* 8.11, repairing at Liverpool to the end of that month.

WELLINGTON alongside the oiler at Moville 7.42, in full war paint. She retains the single 4.7in LA in A and the 3in HA in B, she has 271 certainly and 291 probably. 20mm in the Bridge wings, what seem to be 0.5in quadruple mountings in the bandstands, and HF/DF on its tall mast on the quarterdeck. X 4.7in LA is still mounted.

Following the institution in mid 1941 of the two way convoy system to Freetown, WELLINGTON joined the long distance escorts until 3.42 when she went for an eight week refit at Belfast, to return to the Freetown route on completion. In 11.42 WELLINGTON became involved in the escort of the fast troop convoys (KMF/MKF) to and from North Africa after Operation Torch. This duty lasted until 7.43 when the ship transferred to West Africa Command based at Freetown for duty with the convoy system between Freetown and Nigerian ports.

A respite from the rigours of service on the West coast came in 5.44 when WELLINGTON was sent to Bermuda for refit, remaining there until the end of 7.44, then to return to the Clyde for further work.

WSPL

WELLINGTON in her final form as a Livery Hall in Kings Reach, City of London. Her wartime form, sans armament, is remarkably complete externally.

Returning to Freetown in 9.44, WELLINGTON transferred to 55 EG and the Gibraltar base in 1.45 to serve the remainder of the war there, accepting the surrender of U 541 on 12.5. WELLINGTON returned to the UK in 6.45, and after de-storing arrived at Milford Haven to go into reserve 6.8.45.

On 6.2.47 WELLINGTON was purchased by the Honourable Company of Master Mariners for conversion into a floating Livery Hall. The work involved gutting much of the ship to provide suitable accommodation, in particular the total removal of the boilers and engines, the resultant large space becoming the Livery Hall itself, complete with fine wood panelling, decorative skylight and sweeping timber staircase, all salvaged from passenger vessels being broken up. The ship still remains at her permanent moorings in the Kings Reach of the Thames, a docking in late 1991 ensuring her continued existence for many years to come.

While greatly changed internally, her external appearance is little altered from the original other than for the removal of armament, and she remains the only post 1918 sloop hull extant in the UK pending the possible return of WHIMBREL from Egypt.

RAN GRIMSBY class

Name	Builder	Laid down	Launched	Completed
PARRAMATTA	Cockatoo	9.11.38	18.6.39	8.4.40
SWAN	Cockatoo	1.5.35	28.3.36	10.12.36
WARREGO	Cockatoo	10.5.39	10.2.40	21.8.40
YARRA	Cockatoo	24.5.34	28.3.35	19.12.35

Displacement 1,060 tons. Dimensions length 250ft pp, beam 36ft, draught 10ft max at standard displacement.

Machinery two shaft, geared turbines, designed SHP 2,000 = 16.5 kts. Oil 300 tons, consumption 0.5 tons per hour at 10 kts.

Initial armament. Due to the class's protracted building time, initial and subsequent armaments varied. See table below for individual ships.

PARRAMATTA one twin and one single 4in HA, one quadruple 0.5in mg.

PARRAMATTA as completed. She does not yet mount anything in B position, nor is there any sign of close range AA armament. She carries both M/S gear and two DCTs aft. Note that at this stage she has no rangefinder at the back of the bridge, merely an empty position.

SWAN three single 4in HA, one quad 0.5in mg. Refitted in 1942 with two twin 4in HA, a single 40mm and six single 20mm. Postwar close range armament became four single 40mm, subsequently (as a Training Ship) armament reduced to one twin 4in and a single 40mm.

WARREGO completed with one twin and one single 4in HA, one quadruple 0.5in mg. Refitted 1942 with two twin 4in HA, one single 40mm and six single 20mm. Post war reduced to one twin 4in and four single 40mm, then to one twin 4in and one 40mm, and finally disarmed as a Survey Vessel.

YARRA completed with three single 4in HA and 1 quadruple 0.5in mg, probably still in that state when lost.

HMAS PARRAMATTA L 44, U 44

In a similar manner to many Australian warships, PARRAMATTA was placed at the disposal of the Royal Navy by the Commonwealth government as soon as she had completed her initial training after completion. She followed the logical course of proceeding to the nearest scene of action, then the Red Sea, and in 8.40 assisted in the evacuation of Somaliland then only lightly held by British forces. Subsequently, PARRAMATTA operated as a convoy escort in the Red Sea and, when the Italians had been evicted from their East African territories, went with her sister YARRA to join the Mediterranean Fleet.

During 1941, with the Australian garrison besieged in Tobruk, PARRAMATTA was heavily involved, in company with all the Australian ships in the Fleet and many of the Royal Navy's destroyers and sloops, in escorting convoys to and from the fortress, themselves transporting troops and high value stores. During these activities, she was able to rescue a considerable number of survivors from the sloop AUCKLAND, sunk 24.6.41, but was herself lost when attacked by U 559 on 27.11.41 while escorting an ammunition convoy to Tobruk.

SWAN pre war at Sydney NSW, clearly shows her "GRIMSBY" ancestry, although mounting three single 4in HA.

HMAS SWAN L 74, U 74

SWAN operated in Australian waters after her completion in 1937 and, on the outbreak of war, formed the 20th M/S Flotilla with her sister YARRA and two minesweepers to operate off the eastern seaboard of Australia, it being assumed (correctly) that there was a danger of minelaying by German raiders.

SWAN was still so employed when the war spread to the Pacific in 12.41 with the attacks on Malaya and Pearl Harbor, and shortly afterwards she moved north to counter the threat of the approaching Japanese. SWAN received minor damage in the first air raid on Darwin 19.2.42, and thereafter spent most of her war operating close inshore off New Guinea and amongst the islands of the South West Pacific generally, usually in support of the Australian troops fighting ashore.

When the tide turned against the Japanese, SWAN was in the forefront of amphibious operations, notably at Wewak in 5.45 and had the pleasure of receiving the surrender of the Japanese forces in New Ireland 19.9.45.

Remaining in service as a sloop until 1948, SWAN then paid off and was laid up, to be revived, after refit, in 1956 as a training ship for junior Officers of the RAN, a service that she performed until final decommissioning 21.9.64, the ship being broken up in 1965/66.

An immaculate SWAN dis-armed post war as a Cadet Training Ship. She now has a twin 4in HA forward and a single 40mm in B position, classrooms have been added aft and her masts cut down. A modified radar set tops the Director, otherwise she has only navigational radar.

courtesy R P Hall

WARREGO, probably in 1943 off New Guinea, loaded with Australian troops. Although she has the twin 4in HA in A and X positions, B position has to make do with a quadruple 0.5in mg, although there are single 20mm in the bridge wings. Given the alleged date of the photograph, this seems a very light armament and the photograph may be somewhat earlier. She still has M/S gear aft, and seems only to mount a single DCT on each beam. Note the false bow wave and the unusual punctuation mark between the letter and numerals of the pennant number.

HMAS WARREGO U 73

WARREGO, not completed until 8.40, remained in Australian waters as the mainstay of a minuscule M/S force, and moved north to counter the Japanese threat in early 1942. Like SWAN, she was heavily re-armed to provide some defence against air attack and, when the offensive against the Japanese in the

courtesy K R Macpherson
WARREGO *post war in a somewhat unusual paint scheme. She appears to be serving in the survey role, there are marker poles stowed amidships, and small can buoys on the quarterdeck, while a heavy survey launch is on the davits.*

South West Pacific commenced, she led numerous minesweeper forces during landings, and otherwise operated inshore in support of the Army.

WARREGO was present at the Lingayen Gulf landings (including a brush with two Japanese destroyers), at Mindanao and Panay in early 1945, and finally somewhat further south at the Wewak landings in 5.45.

Post war, WARREGO was dis-armed and employed in the Survey role, remaining on that service (except for refits) until finally paying off 8.8.63. She then lay in reserve, and on the disposal list, until scrapped at the same time as her sister SWAN at Sydney NSW in 1965/66.

courtesy R P Hall
WARREGO *in her final survey role, totally dis-armed and in the usual survey white and buff paintwork, photographed in 1960.*

courtesy R P Hall

YARRA pre war showing her to great advantage. Three single 4in HA in A B and X and 3 pdr saluting guns abaft the searchlight are the total armament, no close range weapons whatever. Probably taken in 1936.

HMAS YARRA L 77, U 77

YARRA, while having a relatively brief wartime career, probably also had the most varied life of the four Australian sloops. Initially, she served in the Red Sea as a convoy escort during the time that the Italians still operated from Eritrea. During this period she was involved in frustrating an abortive destroyer attack on her convoy on 21.10.40; it was perhaps unwise of the Italians to attempt such an action on Trafalgar Day!

In 4.41, still serving on the East Indies Station, YARRA was one of the reinforcements added to an ad hoc troop convoy from India sent to seize Basra following the insurrection in Iraq, the port duly being taken without difficulty on 19.4.41. The problems in the area continuing, it was decided that the occupation of Persia (Iran) could no longer be delayed so YARRA, amongst other ships, took part in Operation Countenance on 25.8.41, the occupation of Abadan and associated ports. While resisted, the operation was a complete success, with only the most minor damage, none to YARRA.

courtesy R P Hall

A very rare view of YARRA in the Persian Gulf 8.41, probably at Basra about the time of her action at Abadan.

Moving to the Mediterranean to join the RAN forces there, YARRA became involved in supplying the fortress of Tobruk and took in two convoys, the second being the final one before the relief of the garrison by the Eighth Army. During this operation the sloop FLAMINGO was heavily damaged aft, and YARRA succeeded in towing the crippled ship to Alexandria 7-10.12.41.

With the threat of war in the East having become reality, YARRA then moved to the Indian Ocean and joined the ad hoc forces operating off the Dutch East Indies, where she undertook the defence of convoys desperately trying to reinforce the area. On the collapse of the Allied defences, the problem then became one of evacuating what was left of the garrisons and the myriad of European evacuees fleeing the advancing Japanese. YARRA's end came on 4.3.42 when, escorting just such a convoy of small craft, she fell in with a force including the Japanese heavy cruisers ATAGO and TAKAO. Hopelessly outnumbered and with no effective means of retaliation against such large opponents, YARRA sank in flames while defending her convoy in a vain attempt to delay its fate.

RIN INDUS class

Name	Builder	Laid down	Launched	Completed
INDUS	H Leslie	8.12.33	24.8.34	15.3.35

Displacement 1,190 tons standard. Dimensions length 296ft 4in, beam 35ft 6in, draught 11ft 4in at full load.

Machinery two shaft, geared turbines, designed SHP 2,000 = 16.25 kts. Oil 341 tons, consumption unknown, but probably approximately 0.6 tons per hour at 10 kts.

Initial armament two 4.7in LA, four 3pdr saluting, one quad 0.5in mg.

Changes during service
No information available; it is likely to have been confined to fitting an additional quadruple 0.5in mg, possibly two single 20mm and the removal of the 3pdrs.

India's second sloop, INDUS, seen here preparing for the Coronation Review of 1937. Internally very similar to the earlier HINDUSTAN, she mounts two 4.7in LA in lieu of the earlier ship's 4in. The forward gun is in B position, which provides a suite beneath it for the Commanding Officer, or Senior Officer when embarked.

HMIS INDUS L 67, U 67

Operating largely off the Indian coast both "showing the flag" and as a training ship, INDUS was placed at the disposal of the Royal Navy on the outbreak of war, to be operated in all respects as a RN ship. Accordingly, she relieved one of HM Ships and was based at Aden from 2.40 to provide a presence there after the usual sloops of the Red Sea Force had been withdrawn for service in Home waters.

When Italy entered the war, INDUS became responsible for escorting the Aden/Suez traffic, bearing in mind the presence of substantial surface and submarine forces in Eritrea and also an efficient air force. This duty continued until the threat receded, when INDUS returned to Bombay to refit throughout 3.41. Returning to the Red Sea, further convoys were escorted and INDUS also took part in Operation

Chronometer, the seizure of Assab by seaborne landing of British and Indian troops from Aden. Thereafter she remained Aden based for general duties until the outbreak of war in the East.

Immediately after the Japanese attack, INDUS was ordered to India to join the East Indies Fleet and generally act in the defence of the sub-continent, as such she arrived at Colombo by the end of the year. In 1.42 she escorted a convoy north to Calcutta, and then the personnel ship *MANOORA* to Rangoon. General duty in support of the retreating Army, along the Burmese coast, followed and it was off Akyab on 6.4.42 that INDUS, with little effective AA weaponry, was attacked and sunk by Japanese aircraft.

BITTERN class

Name	Builder	Laid down	Launched	Completed
BITTERN	White	27.8.36	14.7.37	15.3.38
ENCHANTRESS	J Brown	9.3.34	21.12.34	4.4.35
STORK	Denny	19.6.35	21.4.36	10.9.36

Displacement 1,085 (ENCHANTRESS), 1,100 (STORK), 1,190 (BITTERN). Dimensions length 282ft, beam 37ft, draught 10ft at full load.

Machinery two shaft, geared turbines, designed SHP 3,300 = 18.75 kts. All made 19 kts on trials. Oil 388 tons, consumption 0.66 tons per hour at 10 kts.

Initial armament as designed four single 4.7in, one quad 0.5in mg, but see notes below.

Changes during service
None of the ships completed with the designed armament.

A beautifully clear pre-war view of BITTERN, wearing her Senior Officer funnel band. Very clearly shown is the midships 0.5in mounting, and the clear quarterdeck with no minesweeping winch.

BITTERN, the replacement for the initial ship re-named ENCHANTRESS, completed with a revised armament of three twin 4in HA mountings and a quadruple 0.5in mg. She was lost without any alteration, so far as is known.

ENCHANTRESS, originally named BITTERN, shipped three single 4.7in only, in A, B & Q positions. The latter was removed shortly after completion. On the outbreak of war a single 3 in HA and two quadruple 0.5in mg were added, the latter replaced in 5.43 by four single 20mm. Hedgehog was fitted in 4.42.

STORK completed as an unarmed Survey Vessel, a single 3 pdr saluting gun being carried. When re-armed just prior to the war, she shipped a similar armament to her sister BITTERN. This was augmented by an additional quadruple 0.5in mg, then two single 20mm were added, two further 20mm replaced the quadruple 0.5in in 5.43 and the final war time refit landed all 20mm and fitted three single 40mm Bofors. Hedgehog was fitted in 5.42. Post war, she landed B twin 4in HA mounting when in the Fishery Squadron.

Radar

BITTERN So far as is known, no radar was fitted prior to her loss.

ENCHANTRESS 286 M fitted in 1940, replaced by 286P in 1941. 271 fitted in 6.43, 291 replaced the 286 at an unknown date. HF/DF fitted 4.42.

STORK 286, later replaced by 291, 285 fitted later, 272 fitted in 10.42 and HF/DF in 5.43.

An equally clear if somewhat unfortunate view of BITTERN, taken very shortly before she was sunk by JANUS after serious bomb damage off Norway. Unusually, it shows her still with her L pennant, there had not been time to alter it to U after the 4.40 changes. It seems probable that, nearer home, the ship could have been saved: she floats on an even keel despite the dreadful damage aft.

HMS BITTERN L 07

In fact the second ship of the class, the original BITTERN having been re-named ENCHANTRESS when selected for conversion to the Admiralty Yacht.

The original design, fortunately, having been overtaken by events, BITTERN completed as the first three mounting ship, carrying the new twin 4in HA gun and mounting that was to become the standard secondary armament of the Royal Navy's cruisers, and the main armament of all future sloops. Unfortunately, British industry was incapable of producing a High Angle Control System to match the weaponry, in consequence the effectiveness of the exceptional HA armament of this and subsequent ships was greatly reduced.

The ship, and her sisters, also had another unusual feature; recognising that the motion of a small hull would adversely affect gunnery the Admiralty specified that the recently developed stabilisers should be fitted in an endeavour to reduce this disadvantage. BITTERN therefore carried out extensive "first of class" trials after her commissioning on 15.3.38 with a Portsmouth crew and did not join her Flotilla, the 1st A/S Flotilla of which she was to be Senior Officer ship, until 1.39. Thereafter BITTERN served in the Home Fleet for the remainder of her short peace time career.

On the outbreak of war, BITTERN very logically joined the Rosyth Escort Force, charged with AA escort of East coast convoys. This employment lasted until 4.40 when the Norwegian campaign opened. BITTERN was sent over to take part in Operation Sickle, and then remained on station as a major part of the pitifully small AA defence available in Norway. During the evacuation of Namsos, the ship was hit heavily aft; although still afloat and probably fully capable of tow back to the UK, the imminence of further air attack did not permit such an attempt to be made. Fearful that she would drift inshore into shallower water where salvage of her secret ASDIC equipment might be possible, she was sunk 30.4.40 by torpedo from the destroyer JANUS.

HMS ENCHANTRESS L 56, U 56

The first of the class to be laid down, this ship was to have borne the name BITTERN, but on 7.9.34, prior to her launch, the name was changed to ENCHANTRESS on her selection for completion as the new Admiralty Yacht.

Commissioned at Portsmouth with a crew from that depot on 8.4.35, ENCHANTRESS served as the Admiralty Yacht during the Silver Jubilee Review, and, later, the Coronation Review. Although fitted out as a sloop, and in fact mounting three single 4.7in LA, two forward and one in Q position amidships, she was also fitted with extensive accommodation aft for use by the Board of Admiralty, many furnishings etc being removed from the old ENCHANTRESS prior to the latter's disposal. As the Admiralty Yacht, the ship also wore the traditional livery, black hull with green boot topping, white superstructure and buff funnel, a splendid sight indeed that recalled the old Victorian era.

ENCHANTRESS at the Coronation Review of 1937, following the Royal Yacht through the destroyer lines. Note that she has now landed Q 4.7in, and has a small deck house instead. As the Admiralty Yacht with the Board embarked, she wears the Admiralty Flag, a yellow foul anchor on a red field, at the main.

ENCHANTRESS carried out extensive cruises in the pre-war days, both as the Yacht, as a supplement to the Royal Yacht and also as a flagship for officers normally accommodated ashore. Rear Admiral (Submarines) frequently used the ship when at sea supervising Submarine Flotilla exercises, on other occasions the ship joined local exercises etc. All in all a busy life, with the ship based (when at home) at Portsmouth.

On the outbreak of war, ENCHANTRESS was placed under the orders of Rosyth Command and employed on the East coast as an A/S escort, then being ordered to join Plymouth Command in 10.39. She grounded on Drake's Island on arrival at Plymouth and spent a month in the dockyard repairing. During this time some of her more opulent fittings were removed, and a 3in HA and 2 quad 0.5in mg supplied to give the ship some form of AA protection.

ENCHANTRESS at war, photograph taken in 4.42. She has had much of the extra accommodation aft removed, and mounted 3in HA right aft; quite what is on the bandstand abaft the searchlight is not clear although it could be a 2 pdr in a power mounting. HF/DF is carried on a light pole mainmast; it does not seem as if any radar is yet fitted.

On completion of repairs, ENCHANTRESS joined Western Approaches 1st Escort Vessel Division and operated in the Channel and South Western Approaches. When France surrendered, and convoy traffic of any consequence in the Channel ceased, ENCHANTRESS moved to Liverpool in common with many other escorts, and commenced escorting North Atlantic convoys, arriving from the first one at Sydney CB 21.9.40.

In 12.40 a refit commenced at London, lasting until mid 2.41 and a further lengthy dockyard stay in Liverpool late 5.41-8.41 after which ENCHANTRESS joined 41 EG based at Londonderry for the Freetown route. This activity lasted until 10.42 when she was one of the escorts for the assault convoy, KMF 1, for Operation Torch, the invasion of North Africa, after which she transferred to 61 EG based on Gibraltar as part of the Western Mediterranean Fleet escorting convoys along the north coast of Africa. During this period, on 13.12.42, she took part in the sinking of the Italian submarine CORALLO, damaging herself by ramming the submarine, and returned to Grimsby to refit in 2.43, ramming the dockside on arrival; the subsequent repairs and alterations lasted until 6.43.

Repairs completed, ENCHANTRESS worked up at Tobermory prior to joining 38 EG based at Londonderry from where she went south to Freetown, operating from there until 12.43 when she transferred to the Mediterranean Fleet, still as part of 38 EG. Now Gibraltar based, the routine remained the southern leg of the Freetown convoy route, with a change of Group to 39 EG in 11.44. Four months' refit at Gibraltar 1-4.45 was followed by further escort work, mainly of troop convoys, based on Gibraltar, the ship returning home to refit at Portsmouth in 6.45.

Still possessing much of the accommodation aft originally built for the Board, it was proposed to utilise this and fit out ENCHANTRESS as a Landing Ship Headquarters (Small) for Eastern operations. This work was completed by mid 7.45 and involved the landing of all armament other than her four 20mm, and the installation of communications equipment, rudimentary air conditioning and additional accommodation. ENCHANTRESS was allocated to the Pacific Fleet and sailed for Sydney NSW, reaching Colombo on the day of the Japanese surrender. From Colombo she was ordered direct to Hong Kong and thereafter operated under the British Pacific Fleet control, sailing from Sydney NSW for the UK 27.12.45.

On arrival at Portsmouth in 3.46, ENCHANTRESS was laid up awaiting disposal, and was sold later that year, on 22.10, for £22,500 for commercial service, being renamed *LADY ENCHANTRESS*. Still under this name, and owned by The Three Star Shipping Co Ltd, she arrived in the Tyne 16.2.52 to be broken up, after minimal active commercial service, by Clayton & Davie Ltd at Dunston.

WSPL

ENCHANTRESS post war after disposal. She had been sold 22.10.46 for the quite considerable sum of £22,500 for use as a pleasure steamer operating in the Thames Estuary under the name LADY ENCHANTRESS. The scheme fell through and she was laid up until 1950 when she cruised for a month from Southampton to the Channel Isles. This service also failed, due to defects, and she was again laid up until 1952 at Southampton and then sold for scrap. This view shows her with an extended shelter deck and bridge, and merchant type lifeboats fitted. She appears to be laid up, but whether on the first occasion in the Medway or latterly at Southampton, is not known.

HMS STORK L 81, U 81, F 81

STORK was selected prior to completion for service as a Survey Vessel and was, in consequence, completed "for but not with" her armament. STORK's first survey was in Malayan waters where she arrived at Penang 20.11.36 remaining in that area with occasional visits to Singapore for docking, and also to Trincomalee, until 24.1.39 when she sailed for Devonport.

On arrival, STORK was taken in hand at Devonport, and by mid 9.39, just after the outbreak of war, emerged as a fully equipped AA sloop being ordered to join her sister BITTERN on the East coast based on Rosyth. In 4.40, STORK was ordered to the Norwegian coast; she was with the personnel ship *CHOBRY* when the latter was sunk, was present at the final assault on Narvik, and covered the evacuation of Harstad 7.6.40.

STORK in 1936 as a survey ship. Note the enlarged bridge structure and chartroom aft, also the increased boat stowage. The mainmast is a pole, the apparent tripod is the sounding stave stowage for marking shallow channels.

Returning to the East coast, STORK was bombed 8.9.40 and was under repair at Grangemouth until 18.5.41, after which she worked up at Scapa Flow and joined the Irish Sea Escort Force formed to cover convoys against Luftwaffe attack, then very frequent in the Irish Sea. STORK served in this capacity until the end of 8.41 when she became part of 36 EG based at Londonderry for service escorting convoys south to Freetown and Gibraltar. After completing three convoys to Freetown, STORK was delayed with defects at Gibraltar 20.11-14.12.41, and then sailed with convoy HG 76 under the command of the legendary Commander Walker. On 17.12 she took part in the sinking of U 131, on 19.12 sank U 574, and on 22.12 was badly damaged when rammed by DEPTFORD, all in all a busy convoy! Repaired at Devonport by 20.2.42, STORK returned to her Group and, while with convoy OG 82 on 14.4, took part in the sinking of U 252. Gibraltar convoys remained the ship's usual routine until 9.42, when the Group was used in one of the first Support Group operations, unfortunately soon suspended as the ships were required for the North African invasion, during which STORK escorted the follow-up stores convoy KMS 1.

On 12.11.42 STORK was torpedoed by U 77, and was taken to Gibraltar for temporary repair until 11.1.43; she was then towed (as part of convoy MKS 5) to Falmouth where she remained under repair until mid 7.43.

STORK homeward bound in convoy MKS 5 from Gibraltar 1.43. She lost her bows to a torpedo from U 77 12.11.42. A mounting has been removed to ease top weight forward, and she is on her way home for repair. The weather deteriorated even more than this photograph shows, she and the towing vessel had to leave the convoy.

The refitted STORK in mid 1943, now with a tripod mast. 285 on the Director and 271 aft on a stump lattice tower, HF/DF at the masthead. Hedgehog can be seen to port abaft B mounting, 20mm are in the bridge wings and the bandstands aft.

A Tobermory work up preceded joining 37 EG, again on the Freetown route; STORK showed that even with a new crew, good training and opportunity was still with her when, on 30.8, U 634 fell victim to the escort's attack. STORK remained on the Freetown route until mid 11.43, when she became part of 50 EG of the Mediterranean Fleet, escorting convoys the length of that sea.

STORK returned to Falmouth for a further refit in 2.44 lasting until 20.5, after which there was just time for a work up at Tobermory to prepare the ship for operations in the South West Approaches in support of the Normandy invasion. Once this operation was well under way, the escorts blocking possible submarine attack could be withdrawn, and STORK proceeded to the Mediterranean to join 37 EG in late 7.44. At the end of 11.44 she joined the Gibraltar Escort Force for one month, and then came home for further refit, undertaken at Portsmouth up to the end of 3.45. The inevitable work up completed just in time for the end of the European war, additional work was then needed as STORK was ordered to prepare

STORK in 8.46 serving as Senior Officer of the Fishery Protection Squadron. Note that B mounting has been removed.

for service with the Pacific Fleet, being sent to Newport for the purpose. Here again, the war ended as the post refit work up at Portland did, and STORK paid off into reserve at Portsmouth 17.9.45.

With such a history of extensive repairs so late in the war, it is not surprising that STORK was selected for early post war use, and on 11.1.46 she re-commissioned as Senior Officer ship of the Fishery Protection Squadron in Home waters. She served in this capacity until early in 1948, then reducing to reserve again at Portsmouth where she lay in a high category of reserve, ready for further service, until the end of 1954. By that date, the lack of maintenance and the age of the ship caused her to be downgraded to a lower category of reserve, and she was towed, first to Londonderry, then to Lisahally to lay up until the end of 1957. Listed for disposal, STORK finally arrived at Troon 3.6.58 to be broken up.

EGRET class

Name	Builder	Laid down	Launched	Completed
AUCKLAND	Denny	16.6.37	30.6.38	16.11.38
EGRET	White	21.7.37	31.5.38	10.11.38
PELICAN	Thornycroft	7.9.37	12.9.38	2.3.39

Displacement 1,250 tons standard. Dimensions length 292ft 6in, beam 37ft 6in, draught 11ft at deep load.

Machinery two shaft, geared turbines, designed SHP 3,600 = 19.25 kts. Oil fuel 390 tons, consumption 0.66 tons per hour at 10 kts.

Initial armament four twin 4in HA, one quadruple 0.5in mg, two 3pdr saluting.

Changes during service

AUCKLAND Due to her early loss, no details are known of any changes, which are likely to have been restricted to ad hoc additions to close range AA armament.

National Maritime Museum Negative N31357
An early war view of AUCKLAND clearly showing her "as built" configuration. Note that no radar or close range weapons have yet been fitted, nor are her pennants displayed.

EGRET added, initially, a further quadruple 0.5in mg later supplemented by a single 2pdr. In 5.42 this latter was exchanged for two single 20mm, in 6.43 Hedgehog was added.

PELICAN Initially, a further quadruple 0.5in mg added, later supplemented by two single 20mm, increased in 10.42 to four 20mm. Major refit, in 2.44, amended the armament to three twin 4in HA, a quadruple 2pdr, four single 20mm and Hedgehog.

Radar

AUCKLAND Due to her early loss and Mediterranean service, it is unlikely that any radar was ever fitted.

EGRET 285 and 286 fitted early in 1941, 271 and HF/DF fitted 7.42.

PELICAN 286 fitted early in 1941, supplemented in 2.43 with 271 and HF/DF, 291 replaced 286 during 1943 also.

HMS AUCKLAND L 61, U 61

Originally intended to complete as an unarmed Survey Vessel for service on the New Zealand Station, the allocated name of HERON was accordingly changed to AUCKLAND, regarded as more appropriate. In the event, the international situation cancelled both intentions, and the ship completed as a fully armed escort vessel and relieved LONDONDERRY on the African Station.

Commissioned with a Portsmouth crew on 19.11.38, AUCKLAND worked up at Portland and sailed for her first assignment on 4.1.39, arriving at Simonstown 17.2 and cruised on the eastern coast of southern Africa during that summer. On the outbreak of war she remained Simonstown-based until 11.39 when she went north to Freetown and escorted convoy SLF 9 to the UK, arriving at Portsmouth 30.11.

After refit at Portsmouth to 1.1.40, AUCKLAND joined the Rosyth Escort Force and protected East coast convoys until 4.40 when she went to Norway, remaining in that campaign until 1.5.40. She was then transferred to the Red Sea Force and passed through the Mediterranean during May 1940, to go straight on to Bombay for a one month's refit, from which she returned to the Red Sea in time to cover the evacuation of Berbera.

AUCKLAND remained on Red Sea convoy duty until 1.41 when she returned to Bombay for a five week docking and refit, after which she went back to Red Sea convoy duty until the beginning of 4.41. On 8.4.41 AUCKLAND arrived at Alexandria and transferred to the Mediterranean Fleet being based at Alexandria and covering convoys and operations along the Libyan coast until sunk by air attack off Tobruk on 24.6.41.

EGRET on 28.11.38 shows the heavy, twin 4in HA armament, the ultimate sloop development. However, the sole close range armament is the puny quadruple 0.5in in Q position, quite inadequate for the coming conflict. Note that the M/S winch has been displaced by Y mounting, and that an adequate DC armament is fitted.

HMS EGRET L 75, U 75

EGRET commissioned with a Portsmouth crew 11.11.38 and, after a Portland work up and Christmas leave, sailed for the Red Sea where she was to become Senior Naval Officer's ship, arriving on station 31.1.39. Following the usual pattern of cruising with senior dignitaries embarked as guests of the SNO, on the outbreak of war EGRET became part of the East Indies Station, the Red Sea designation lapsing, and went south to establish a patrol off Lourenco Marques to intercept German shipping.

In 11.39 EGRET was ordered home, the long way round as it happened, being routed via Aden to Gibraltar, then south to Freetown to collect convoy SLF 13 to the UK, arriving 4.1.40 and promptly colliding with *SEA VALOUR* on arrival. Repairs at Cardiff lasted for three weeks, after which EGRET became part of the Rosyth Escort Force for the whole of 1940.

On 31.1.41 EGRET arrived at Londonderry where she was based for five months escorting Atlantic convoys in and out of the northern approaches to the UK. In 6.41, she came to London for a six weeks' refit, returning to Londonderry and joining the newly formed 44 Escort Group for the Freetown convoy route. Initially, the escorts on this route terminated at Bathurst, and it was here that EGRET grounded on 2.1.42, fortunately without damage.

Service to and from Freetown continued until 10.42, interrupted by a refit on the Tyne 22.5-24.7.42. At the end of 10.42 the Freetown convoys were suspended in preparation for the invasion of North Africa, Operation Torch, and EGRET escorted numerous troop convoys to and from North Africa from 1.11.42 to 4.43, when she went for refit on the Humber.

On completion of refit 6.7.43, EGRET joined 10 EG also based at Londonderry, and resumed escorting troop convoys, taking KMF 20 out in 7.43 and returning with MKF 20. EGRET and her Group were then diverted to take part in the Biscay anti-submarine drive titled Operation Percussion, and fell victim to a new weapon recently introduced into German service. While patrolling in Biscay 27.8.43, EGRET came under air attack and was hit by a glider bomb. This was only the second such attack (the first was on BIDEFORD two days earlier) and EGRET had the misfortune to be the first warship lost by this means. 5 Officers and 30 ratings survived the attack and were picked up by the Canadian destroyer ATHABASKAN, herself damaged during the same attack.

PELICAN at sea during the war. She now ships 285 and 291 radar, oddly no 271 is fitted yet. Single 20mm are in the bridge wings, the original quadruple 0.5in remain amidships, aft there are four DCT and the usual DC rails.

HMS PELICAN L 86, U 86, F 86

PELICAN was the subject of a number of role changes during her construction, being initially intended for service as a survey vessel, then as an escort vessel for the Red Sea, the orders then being amended again to a survey vessel for service in China, and finally to complete as a fully armed escort. Even that was not the end of change, for the decision that she was to serve on the America & W Indies Station caused a manning depot change from Devonport to Chatham, duly commissioning on 2.3.39 at her builders. Twelve days later came a further change of mind, that this modern unit was to remain in Home waters, and that she was to exchange crews with the older PENZANCE so that the allocated crew and the older ship could go overseas, while the modern hull remained at home serving in the Fishery Protection Squadron.

PELICAN thus spent the summer of 1939 cruising off the west coast of Britain and off Iceland, joining the Rosyth Escort Force on the outbreak of war. Her early war service was something of a disaster for the ship; she was in collision with *STARLING* (the freighter, not the warship) 5.9, struck wreckage 20.10 and, during the Norwegian campaign, was bombed on 22.4. This last incident caused PELICAN to repair at Chatham until 5.12, but on Boxing Day she was again in collision, this time with the trawler CAPE PORTLAND requiring further repair work at London during 1.41.

PELICAN can only be described as accident prone, for only ten days after completion of her repairs she was mined on 19.2, and returned to London for further repair work lasting until 1.12.41. Even this was not the end, because en route to work up she suffered minor damage in an air attack, so that it was early 1.42 before PELICAN joined 45 EG at Londonderry for service with the Freetown convoys.

PELICAN 6.7.50 as a Leader in the Mediterranean. She still retains her tripod mast, unusual in sloops at this time. X 4in mounting has been removed in favour of quadruple 2 pdr, and single 20mm are mounted amidships and in the bridge wings.

PELICAN escorted shipping to Freetown until the suspension of that route in 10.42; while with the outward convoy OS 33 on 11.7 she took part in the sinking of U 136. PELICAN escorted the second troop convoy for Operation Torch, and then spent 12.42 under repair at Belfast. During this refit there were at least two separate incidents of sabotage on board when gun sights were deliberately damaged, followed by a further incident in late 1.43 while the ship was in North Africa, when the gyro compass was disabled.

In 2.43 PELICAN joined 1 EG and, after one return passage to Freetown, the Group transferred to the North Atlantic where it operated in the Support Group role from 4.43. Almost at once she was successful when, on 6.5, U 438 was sunk, followed by a second "kill" on 14.6, when U 334 was destroyed. The Group returned to normal escort duty in 7.43 with further troop convoys to North Africa.

In 8.43 1 EG came under Devonport Command, but remained responsible for major North African troop convoys, PELICAN continuing this work until going for refit at Milford Haven 12.12 to the end of 2.44. After post refit work up PELICAN joined the Biscay anti-submarine patrols, resulting in the sinking of U 448 on 14.4. A brief repair period, and the Group became one part of the many A/S Groups charged with sealing the approaches to the Normandy invasion area against submarines, work which occupied PELICAN until 31.7 when she went to the Clyde for a six week refit.

WSPL
Still in reasonable external appearance, but dis-armed, PELICAN lies at Preston awaiting the breakers' torch having arrived at the yard 29.11.58.

On completion of refit she was ordered to join the Eastern Fleet and arrived at Bombay 5.10.44 in need of repair, having grounded at Aden on 2.9. Repairs lasted until 17.11 when it became apparent that further work was required that could not be undertaken at Bombay. PELICAN was therefore sent to the Italian Dockyard at Taranto, then under British control, where she remained until 3.8.45. On completion, further work was carried out at Malta to 12.9, PELICAN then joining the Mediterranean Fleet where she was to serve continuously until 1951.

Her Mediterranean service was in 33 EG to mid 46, then as Senior Officer of 5 Escort Flotilla to 10.47 when she transferred to 2 Escort Flotilla as Senior Officer until 6.51 when she paid off to reserve at Chatham. Even this was not the end of her service; she re-commissioned in 8.54 after refit for service on the South Atlantic Station, remaining in commission until 13.2.57 when she came home and paid off for the last time. Laid up again, she finally arrived at Preston on 29.11.58 for breaking up by T W Ward Ltd.

RIN Modified BITTERN class

Name	Builder	Laid down	Launched	Completed
GODAVARI	Thornycroft	31.10.40	21.1.43	28.6.43
JUMNA	Denny	20.2.40	16.11.40	13.5.41
NARBADA	Thornycroft	30.8.41	21.11.42	29.4.43
SUTLEJ	Denny	4.1.40	1.10.40	23.4.41

Displacement 1,300 tons standard (JUMNA & SUTLEJ) 1,340 tons. Dimensions length 292ft 6in, beam 37ft 6in, draught 11ft (JUMNA and SUTLEJ only) 9ft.

Machinery two shaft, geared turbines, designed SHP 3,600 = 19 kts on trials. Oil 390 tons, consumption 0.66 tons per hour at 10 kts.

Initial armament three twin 4in HA, two single 2pdr, one twin 20mm, two single 20mm in GODAVARI and NARBADA; three twin 4in, one quadruple 0.5in, two single 20mm in JUMNA and SUTLEJ.

Changes during service

GODAVARI Added one twin 20mm during wartime service.
JUMNA Added four single 20mm in lieu of 0.5in.
NARBADA No reported changes.
SUTLEJ Added four single 20mm in lieu of 0.5in, added Hedgehog.
Radar
GODAVARI 285, 271 and 291 radars had been fitted initially.
JUMNA 285 and 286 radar had been fitted initially.
NARBADA 285, 271 and 291 radars had been fitted initially.
SUTLEJ 285 and 286 radars fitted initially.

A smart GODAVARI just out of the builders 25.6.43, clearly showing her 40mm amidships. As in so many photographs, she has trained her armament at the camera, and for good measure added the Director as well!

HMIS GODAVARI U 52, F 52

GODAVARI spent five weeks working up at Scapa Flow, this venue reflecting her probable service in the AA rather than the A/S role. On completion she operated for some time in Home waters, including escorting convoys to and from Gibraltar; she was also in collision with MANCHESTER PROGRESS on 22.12.43 and repaired at Londonderry. GODAVARI was next ordered to join the Eastern Fleet, then based at Trincomalee, and accordingly escorted two convoys outward to Port Said, and a further convoy from Port Said to Bombay where she finally arrived in her home country 25.4.44.

GODAVARI was then employed extensively on convoy escort and patrol duty in the western Indian Ocean, participating in several offensive sweeps to locate long range U boats operating there. On one such operation, involving patrols from Trincomalee to Kilindini as part of Force 66, U 198 was located and sunk 12.8.44.

On the Fleet re-organisation of 11.44, GODAVARI became part of the East Indies Fleet; shortly afterwards she went to refit at Bombay 26.12.44 to 28.4.45, thereafter being based on Ceylon until the end of the war. In 8.45, she was one of the ships ordered to prepare for Operation Zipper, an assault landing on Penang and Port Swettenham; the Japanese surrender taking place prior to this assault, the delayed operation instead became the re-occupation of Malaya and Singapore in which GODAVARI took a full part, eventually arriving in Hong Kong 20.10.45.

Post 1947, GODAVARI became the Pakistan SIND, and under that flag cruised world wide, usually in the training ship role. Here, still fully armed, she turns in Sydney Harbour during an official visit to Australia.

Post war, when the partition of the old Indian state took place, GODAVARI was allocated to Pakistan under the division of assets and duly became SIND of the new Royal Pakistan Navy. After considerable service she was finally sold 2.6.59 for breaking up, presumably being scrapped in Pakistan.

HMIS JUMNA U 21, F 21

JUMNA completed her work up 29.6.41 and, after rectification of defects, operated for some weeks between Milford Haven and Belfast as AA defence for convoys on that route, the Luftwaffe being particularly active in the area. In 9.41, she was ordered to start her passage to the East and escorted convoy OS 6 to Freetown on the first leg of her long passage via Capetown. From Capetown she was ordered to Suez where she arrived 18.11.

When war spread to the Pacific in 12.41, JUMNA was ordered to Indian waters and at the end of the year escorted a troop convoy, DM 1, with reinforcements for Singapore, and arrived at that base 13.1.42. She then operated locally, and finally left Singapore 25.1.42 with a convoy to Colombo, thus avoiding the collapse of the fortress three weeks later.

JUMNA refitted at Bombay 13.3-25.4 and then escorted Bombay/Colombo convoys until 8.42 when she moved to the Bombay to Aden and Persian Gulf routes on which duty she remained, with little variation, until 5.43.

In 5.43 JUMNA was lent to the Mediterranean Fleet and operated there until mid 9.43, a period of activity that saw the invasion of Sicily and the collapse of Italy. At the end of 9.43 JUMNA was ordered home and between then and the end of the year escorted convoys to and from Aden and Kilindini, based on Bombay. After refit at Bombay 22.12.43-31.1.44 JUMNA moved to Calcutta escorting convoys from Bombay to Trincomalee and then to Calcutta while on passage. During the escort of convoy JC 36 she located and sank the Japanese submarine Ro 110 on 11.2.

The first half of 1944 was taken up with escort duty, principally on the Calcutta/Chittagong route which was the supply line for the Army in Burma. On 30.6 JUMNA commenced a refit in Bombay which lasted until 21.9 and then returned to convoy escort and patrol work.

courtesy K R Macpherson
JUMNA at sea 4.10.44, photographed in the Indian Ocean south of Ceylon, looking almost post war in her uncamouflaged smartness.

At the end of 1944 JUMNA took part in Operation Lightning, and then became closely involved in operations along the Burmese coastline, service that continued until 6.45. The ship then needing docking and refit, and the Indian yards being somewhat overloaded, JUMNA was despatched to the Red Sea port of Massawa where the former Italian dockyard was operated by the Royal Navy with a good deal of US expertise added. This refit lasted until 23.11.45 after which JUMNA returned to Bombay.

JUMNA remained in Indian service after the partition of assets in 1947, the English transliteration of her name eventually becoming JAMUNA. She was employed, after service as a sloop - later frigate, as a Training Ship and also for survey duties, finally being paid off for disposal at the end of 1980.

JUMNA, or more properly JAMUNA in the later transliteration of her name, as a survey ship in Indian service in 1970. She was to serve for a further ten years in this role before paying off finally at the end of 1980.

The Indian "Modified BITTERN" NARBADA shortly after completion.

Fotoflite incorporating Skyfotos
In 1947, NARBADA passed to Pakistan as JHELUM; here she is seen off Portsmouth having just completed the finishing touches to her paintwork before going up harbour to Spithead to join the Coronation Review in 6.53. She now carries single 40mm amidships, but retains her old 20mm in the bridge wings. Radar remains the wartime 285 and 291 plus a navigational set on the pole mainmast.

HMIS NARBADA U 40, F 40

NARBADA worked up at Tobermory and Scapa Flow, completing post work up defects 12.8.43 and escorting a troop convoy (KMF 22) to Algiers shortly afterwards. From Algiers, she escorted a freight convoy to Alexandria, and then a further convoy from Aden to Bombay, arriving at her home base for the first time on 14.9 to become part of the Eastern Fleet.

By late 10.43 NARBADA was working the supply routes to and from Chittagong in support of the Army on the Indo-Burmese border, operations that continued almost unbroken until 1.45. NARBADA then switched to active support of the numerous landings and support operations along the Burmese coast, where sloops became the "battleships" of the minor forces harrying the retreating Japanese, often far upstream from the open sea. This duty continued until 3.45, when NARBADA was withdrawn for an overdue refit, arriving 16.3 at Bombay where she was in dockyard hands until 23.5.

NARBADA then returned to the Trincomalee base, and operated from there for the remainder of the Japanese war, visiting Rangoon 14.9 and arriving at Port Blair 26.9 for the Japanese surrender there.

Serving in the RIN, NARBADA passed on the partition of India to Pakistan, being commissioned as HMPS JHELUM for further service. She was ultimately sold 15.7.59 and broken up in Pakistan.

SUTLEJ, probably at Bombay, in late 1943. Only 285 and 291 are mounted, and a light 20mm armament.

HMIS SUTLEJ U 95, F 95

SUTLEJ had a long work up both at Tobermory and Scapa Flow, finally completing post work up defects 25.6.41. She then operated between Milford Haven and Belfast as AA defence for convoys on that route until 8.41 and was then ordered to India via Capetown and Aden.

SUTLEJ made her passage via Freetown, St Helena, Durban and Mombasa, finally arriving 4.11.41 at Suez to become an AA guardship in the roads there, until the start of the war in the Pacific.

Ordered to Ceylon when the Japanese attacked, SUTLEJ sailed from Suez 11.12 and arrived at Colombo 23.12 to start escort duty from that base. She took two troop convoys through to Singapore (BM 10 and BM 12) during the vain attempts to bolster the defences there, and was fortunate to sail on 7.2.42 from Singapore with orders to proceed to Trincomalee, thus avoiding the fate of many of the ships which left later. During 3 to 5.42 SUTLEJ was engaged on the east coast of India convoy routes, in late 5.42 she shifted her operations to the west coast, and in 9.42 to the Bombay to Arabian Sea routes, in which area she remained active until 5.43 when she was sent to the Mediterranean.

In the Mediterranean, initially with the Mediterranean Fleet and from 7.43 with 1 Convoy Escort Group of Levant Command, she escorted convoys along the Libyan and Tunisian coast supplying the Eighth Army in its westward advance. After the occupation of Sicily, SUTLEJ moved to a fresh area of operations, escorting convoys northwards from Alexandria to Palestine and the Lebanese coast until the end of November.

On 1.12 SUTLEJ sailed for Bombay, arriving there with a convoy from Aden 27.12, then to go further east and take up escort work between Calcutta and Chittagong in late 3.44. She remained on this duty until the end of 6.44 when she became based in Ceylon for three months engaged on A/S patrols from Trincomalee.

At the end of 9.44 SUTLEJ went to Bombay for a long refit, not completing until 7.3.45 when she returned to the Ceylon area in time to take part in Operation Dracula, the seaborne occupation of

SUTLEJ well post war, serving as a survey ship in the Indian Navy.

Rangoon, in 4.45. Further patrols off the Andamans followed, but defects developed and SUTLEJ returned to Bombay for a refit 3.8.45. The end of the Japanese war, and turmoil in Indian circles generally, delayed the completion of this so that it was early 2.46 before the ship left dockyard hands.

SUTLEJ remained in Indian service after the partition of the country and, after long service as a frigate, was converted to the survey role, finally paying off for disposal at the end of 1978. She was presumably broken up in India, but unfortunately no details are available.

BLACK SWAN class

Name	Builder	Laid down	Launched	Completed
BLACK SWAN	Yarrow	20.6.38	7.7.39	27.1.40
ERNE	Furness SB/R Westgarth	22.9.39	5.8.40	26.4.41
FLAMINGO	Yarrow	26.5.38	18.4.39	3.11.39
IBIS	Furness SB/R Westgarth	22.9.39	28.11.40	30.8.41

Displacement 1,300 tons. Dimensions length 299ft 6in oa, beam 37ft 6in, draught 11ft 6in max at standard displacement.

Machinery two shaft, geared turbines. SHP 4,300 = 20 knots. Oil 428 tons, consumption 0.75 tons per hour at 10 knots.

Initial armament as designed three twin 4in HA, two quad 0.5in mg.

Changes during service
BLACK SWAN Shipped a quadruple 2pdr in 5.41 and exchanged her 0.5in for single 20mm in 9.41. Added two further 20mm and Hedgehog 6.42, increased 20mm to six in 9.43, and in 5.45 amended 20mm suite to two twin and two single weapons.

ERNE Completed with a quadruple 2pdr and two single 20mm, increased to seven single 20mm and fitted Hedgehog 10.42. Post Op Torch reduced 20mm to the standard six weapons.

FLAMINGO Completed with her quadruple 2pdr, then added two single 20mm and Hedgehog, and added a further two single 20mm later.

IBIS Completed with quadruple 2 pdr and two single 20mm, no known changes prior to loss though possibly added two single 20mm for Op Torch.

Radar
BLACK SWAN Fitted 286 in early 1941, and HF/DF in 9.41; added 271 7.42, 273 and 285 in late 1943. ASDIC upgraded to 147S 9.43.

ERNE 285 and 286 fitted 7.41, HF/DF 12.41 and 271 8.42.

FLAMINGO No information available, but unlikely to have fitted radar prior to Bombay refit.

IBIS Known to have fitted HF/DF 11.41, no radar information available.

HMS BLACK SWAN L 57, U 57, F 57

Completed 27.1.40, BLACK SWAN worked up at Portland during 2.40 and then joined the Rosyth Escort Force for East coast duty until the start of the Norwegian campaign, when she became one of the AA guardships endeavouring to protect the landings there. During this duty she was bombed on 27.4, the bomb fortunately passing through the magazine and out of the ship before exploding; the damage was repaired at Falmouth during 5.40.

Returning to East coast escort work, BLACK SWAN was mined 1.11 and repaired at Dundee until 16.5.41, thereafter working up at Scapa Flow and joining 1st AA Division of Western Approaches based on Belfast, on 27.6. Duty with this unit entailed escorting the convoys from Milford Haven, the collecting point for Bristol Channel ports, to Belfast for onward routing. The Luftwaffe expended a good deal of effort attempting to interfere with this traffic, and heavy AA escort was called for.

BLACK SWAN was bombed, with minor damage, at Milford Haven 25.8, repairing at Pembroke Dock to 11.9. In 10.41 she was transferred to Londonderry for ocean escort work, and joined 37 EG based there at the end of 11.41 being employed principally on the Gibraltar route. BLACK SWAN was involved in

courtesy W A Fuller

BLACK SWAN seen in 9.42. Note the 271 on the stump tower aft, the 285 aerials on the Director and the HF/DF aerial at the masthead. The eight DCTs are clearly shown, and there is a substantial stowage of smoke floats on top of the DC rails aft.

Operation Torch, and continued escorting Gibraltar bound trade until 3.43 when she transferred to the Freetown route. On 2.4.43, the ship took part in the sinking of U 124.

In 6.43 BLACK SWAN incurred damage, either from a premature depth charge or from the explosion close alongside of a torpedo; whatever the cause, she had to repair at Liverpool from 7.6-8.10, followed by work up at Tobermory. Defects occurred during work up, and further repair was needed at Troon for six weeks, following which BLACK SWAN was ordered to the Mediterranean for convoy work there.

BLACK SWAN proceeded to the Mediterranean escorting convoy KMF 28, and joined 51 EG to operate with convoys throughout the Command until the end of 9.44 when she returned to Western Approaches. Based on Holyhead, and then the Clyde, her employment was brief as she went for refit at Leith from 1.12.44-9.4.45 prior to joining the Pacific Fleet.

After a Tobermory work up, BLACK SWAN escorted one convoy in UK waters, and then had a six week defect programme at Devonport before sailing for Malta 22.6. At Malta, she carried out further training until mid 7.45, and then sailed for the Pacific, arriving at the fleet base of Manus early in 9.45. Japan having surrendered, BLACK SWAN's first duty was to proceed to Hong Kong as part of the relieving squadron, thereafter remaining with the Pacific Fleet as part of the Escort Force.

BLACK SWAN late in the war wearing Pacific camouflage. Note the Hedgehog abaft B gun, the new lattice mast with a strengthening stanchion for the 293 at yardarm level. 20mm are mounted in the bridge wings and the bandstands, and her four DCT fit is clearly shown.

In 1.47 BLACK SWAN became part of 1st Escort Flotilla as Senior Officer, transferring in early 1949 to 3rd Escort Flotilla, in which she remained until returning to the UK and paying off to reserve at Portsmouth in 5.52. Transferred to Devonport by the end of the year, BLACK SWAN then shifted again to Lisahally in mid 53 where she remained until towed to Troon for scrapping, arriving there 13.9.56.

ERNE, in a somewhat dilapidated state. This photograph can also be found with the pennant altered to read 08, but compare this view with that of WOODPECKER on a later page. Here the 285 and HF/DF aerials are very clear, there are single 20mm in the bridge wings and bandstands, and a quadruple 2 pdr on the quarterdeck. The masthead carries 291, and the 271 tower is unusually tall. By comparison, WOODPECKER has a shorter 271 tower, 291 at the masthead, 2 pdrs in the bandstands and 20mm on the quarterdeck, see page 112.

HMS ERNE U 03

ERNE had an inauspicious start to her life, receiving severe damage from a near miss 30.4.41 while still at her builders yard, four days after completion. The resultant damage sent her to the Tyne for permanent repair, which lasted until 21.6.42 followed by work up at Tobermory before joining 45 EG in 7.42, fifteen months after her first commissioning.

Early success was achieved when ERNE took part in the sinking of U 213 on 31.7 while with convoy OS 35 en route to Freetown. She remained on this route until Operation Torch, when she transferred to 43 EG of Western Mediterranean Command in 11.42, remaining in that force until 2.43 when she came home to Londonderry and joined 44 EG. From that base, ERNE escorted convoys on the Gibraltar route, being in collision with *EMPIRE MINNOW* while at Gibraltar on 14.4.

In 7.43 came a change of scene when ERNE escorted the US convoy GUS 10X from Gibraltar to Norfolk Va, returning to the UK via Halifax with HX 255, on arrival going to refit at Belfast 12.9.43-1.1.44, joining 10 EG based on Londonderry on completion. After a short period with 10 EG, ERNE shifted to the Mediterranean, and a protracted repair period at Taranto to 21.9, the cause of which remains unknown. Whatever the cause, work done at Taranto could not have been particularly satisfactory for further repairs were undertaken at Alexandria during 12.44 after which ERNE sailed for Malta.

Grounding on arrival at Malta, ERNE was sent back to the UK and spent two months repairing at Falmouth, prior to a Tobermory work up during 4.45. The ship then worked out of Portsmouth for the last days of the European war, and then sailed for the Pacific at the end of 6.45.

ERNE spent a month at Malta exercising and joined the Pacific Fleet, passing through Colombo 11.8 and being diverted direct to Tokyo to join the main body. She served only briefly in the Pacific, returning to Devonport 19.2.46 and paying off to reserve.

It must be assumed that her unfortunate start to life affected the main structure of the ship, hence her frequent repair periods, and also her early disposal. Languishing in a low category of reserve, ERNE was selected in 1950 to be converted to a static drill ship for Solent Division RNVR.

ERNE in her final guise as the RNVR drillship WESSEX, in fact towed out for the Coronation Review of 1953. She has been stripped, both externally and internally, and she displays only instructional radar sets, no armament and minimal boats.

Stripped and fitted out at Portsmouth, ERNE took up her new duties 4.6.52, changing her name to WESSEX, the name borne by Solent Division. She remained in this role alongside at Southampton until new headquarters were acquired ashore by the Division, she was then sold on 27.10.65 for breaking up, work which was carried out at Antwerp during 1966.

HMS FLAMINGO L 18, U 18, F 18

FLAMINGO, on completion, joined the Rosyth Escort Force and was almost immediately in collision with *DOWNLEAZE* in convoy FN 137 on 16.11.39. A month's repair at Leith followed prior to her return to East Coast duty.

In 4.40, like most of the modern sloops, FLAMINGO went to the Norwegian campaign and received near miss damage 30.6.41 which disabled her and she was towed to Rosyth by the cruiser CURACAO. Repaired by mid 5.40, she was allocated to the Red Sea Force and arrived at Port Sudan via the Mediterranean 23.5.40, operating principally in the Red Sea until early 4.41 when she transferred to the Mediterranean Fleet to take part in Operation Demon, the landing of the British Army in Greece. Thereafter she was based at Alexandria and operated mainly to and from Tobruk during the siege until heavily bombed 7.12.41, receiving severe damage aft.

A photograph of FLAMINGO, presumably taken in the Mediterranean from the tropical uniform; she has an Admiral's flag at the foremast. There are no radars nor any sign of close range weapons other than the 2 pdrs, which indicate an early 1941 date in all probability.

FLAMINGO, unrepaired, was laid up in Suez Roads as a static AA guardship; due to lack of maintenance her damaged hull deteriorated until she was unable to fire the main armament due to her advanced state of corrosion. Finally she departed Suez 5.2.43 in tow of the tug *AKBAR* to Aden, arriving 15.3, and onward from Aden in tow of *STAR OF CAIRO* to Bombay. Arriving 26.3.43, she remained under repair at Bombay until 3.1.44 when she re-commissioned for trials, joining the Eastern Fleet 16.1.44 on satisfactory completion.

In 6.44 FLAMINGO became part of 61 EG of the Eastern Fleet, based principally on Colombo, remaining there with the East Indies Fleet on the re-organisation in 11.44.

Sailing for the UK 20.4.45, FLAMINGO arrived in Liverpool 13.5.45 and then refitted prior to allocation to the British Pacific Fleet. The sudden end to the Pacific War cancelled these plans, and FLAMINGO was laid up in Reserve at Devonport until the end of 1948.

WSPL

The post war FLAMINGO. Note the extra cabin block on the quarterdeck, 291 on the stump aerial spreader, single 40mm in the bridge wings, and a new lattice mast.

Re-commissioned in 1.49 FLAMINGO went to the Persian Gulf Division, returning to Chatham to refit in late 1952. By 1.54 she was back on station in the Gulf returning to Devonport to lay up in reserve in late 1955. Scheduled for disposal, she was sold to the Bundesmarine in late 1957 and, after refit, was commissioned as the GRAF SPEE 21.1.59.

courtesy Bundesmarine
The German GRAF SPEE in 1963, the former FLAMINGO. The classic sloop hull remains unaltered as does the bridge, only the distinctive armament, funnel cap and classroom block aft have altered her from the previous photographs.

A late 1941 view of IBIS in a somewhat surreal camouflage style, the foreman painter at the builder's must have enjoyed himself! As befits the early date, a very basic fitting, 285 is the only radar but she does have two single 20mm in the bandstands by the funnel to supplement the quadruple 2 pdr on the quarterdeck.

HMS IBIS L 99, U 99

On completion of work up IBIS joined 41 EG based at Londonderry and escorted convoys to and from Freetown in the OS/SL cycle from 11.41 to 7.42, her final passage being with SL 116 when she brought home, in addition to the convoy, a substantial amount of gold bullion. IBIS then went to refit at Tilbury from 10.8.42 to 23.9.42 to prepare for the forthcoming invasion of North Africa.

Prior to Operation Torch, IBIS escorted convoy KX 1, one of the build up convoys for Gibraltar, and then returned to the actual landings as part of the escort of the assault convoy KMF 1. Two days after the landings, on 10.11.42, she was sunk by Italian torpedo bombers north of Algiers.

Modified BLACK SWAN class

Name	Builder	Laid down	Launched	Completed
ACTAEON	Thornycroft	15.5.44	25.7.45	24.7.46
ALACRITY	Denny	5.4.43	1.9.44	13.4.45
AMETHYST	Stephen	25.3.42	27.5.43	2.11.43
CHANTICLEER	Denny	13.6.41	24.9.42	29.3.43
CRANE	Denny/J Brown	13.6.41	9.11.42	10.5.43
CYGNET	C Laird	30.8.41	28.7.42	1.12.42
HART	Stephen	27.3.42	7.7.43	12.12.43
HIND	Denny/J Brown	31.8.42	30.9.43	11.4.44
KITE	C Laird	25.9.41	13.10.42	1.3.43
LAPWING	Scott's	17.12.41	16.7.43	21.3.44
LARK	Scott's	5.5.42	28.8.43	10.4.44
MAGPIE	Thornycroft	30.12.41	24.3.43	30.8.43
MERMAID	Denny	8.9.42	11.11.43	12.5.44
MODESTE	Chatham/Yarrow	15.2.43	29.1.44	3.9.45
NEREIDE	Chatham/Yarrow	15.2.43	29.1.44	3.9.45
OPOSSUM	Denny	28.7.43	30.11.44	16.6.45
PEACOCK	Thornycroft	29.11.42	11.12.43	10.5.44
PHEASANT	Yarrow	27.3.41	21.12.42	12.5.43
REDPOLE	Yarrow	18.5.42	25.2.43	24.6.43
SNIPE	Denny	21.9.44	20.12.45	9.9.46
SPARROW	Denny	30.11.44	18.2.46	16.12.46
STARLING	Fairfield	21.10.41	14.10.42	1.4.43
WHIMBREL	Yarrow	31.10.41	25.8.42	12.1.43
WILD GOOSE	Yarrow	28.1.42	14.10.42	11.3.43
WOODCOCK	Fairfield	21.10.41	26.11.42	29.5.43
WOODPECKER	Denny	23.2.41	29.6.42	14.12.42
WREN	Denny	27.2.41	11.8.42	4.2.43

Displacement 1,925 tons standard. Dimensions length 299ft 6in, beam 38ft 6in, draught 11ft 4in at deep load.

Machinery two shaft, geared turbines, designed SHP 4,300 = 19.75 kts. Oil fuel 390 tons, consumption 0.8 tons per hour at 10 kts.

Initial armament three twin 4in HA in all. Close range armament varied on completion between ships according to date, see below for individual details.

Variations in initial close range armament, and subsequent changes

ACTAEON Completed with two twin 40mm and four single 20mm. Post war two twin and two single 40mm.

ALACRITY Completed with two twin 40mm and two single 20mm. Post war two twin and two single 40mm.

AMETHYST Completed with four twin and four single 20mm, and Hedgehog. Refitted 5.45 with two twin 40mm, two twin and two single 20mm. Post war two twin and two single 40mm, and Hedgehog.

CHANTICLEER Completed with four twin and two single 20mm, and Hedgehog. Disarmed 12.43 when hulked.

CRANE Completed with four twin and two single 20mm, re-armed with two twin 40mm, one twin and two single 20mm. Post war two twin and four single 40mm, and Hedgehog.

CYGNET Completed with two quad 2pdr and four single 20mm, refitted 5.45 with two twin 40mm and four single 20mm.

HART Completed with two twin and two single 20mm, and Hedgehog. Refitted with two twin 40mm, two twin and two single 20mm. Post war two twin and two single 40mm.

HIND Completed with four twin and two single 20mm and Hedgehog, refitted with two single 40mm, four twin and two single 20mm. Post war two single 40mm, two single 20mm.

KITE Completed with 2 quadruple 2pdr, four twin and two single 20mm.

LAPWING Completed with four twin and two single 20mm.

LARK Completed with four twin and two single 20mm.

MAGPIE Completed with four twin and four single 20mm, and Hedgehog. Refitted with two twin 40mm, two twin and two single 20mm. Post war two twin and two single 40mm.

MERMAID Completed with four twin and two single 20mm, and Hedgehog. Refitted with two twin 40mm, four twin and two single 20mm. Post war two twin and two single 40mm.

MODESTE Completed with two twin and two single 40mm, added two single 40mm post war. Final armament two twin and three single 40mm and Hedgehog.

NEREIDE Completed with two twin 40mm. Post war two twin and two single 40mm.

OPOSSUM Completed with two twin 40mm, two twin and two single 20mm. Post war two twin and four single 40mm, and Hedgehog.

PEACOCK Completed with four twin and two single 20mm, refitted with two twin 40mm, four twin and two single 20mm. Post war two twin and two single 40mm, and Hedgehog.

PHEASANT Completed with four twin and two single 20mm, and Hedgehog. Refitted with two twin 40mm additional. Post war two twin and two single 40mm.

REDPOLE Completed with two twin 40mm, two twin and two single 20mm, and Hedgehog. Refitted with two twin and two single 40mm, one twin and two single 20mm. Disarmed post war.

SNIPE Completed with two twin and two single 40mm, and Hedgehog, retaining this suite throughout.

SPARROW Completed with two twin and four single 40mm.

STARLING Completed with four twin and two single 20mm, refitted with two twin and two single 40mm, one twin and four single 20mm. Disarmed post war.

WHIMBREL Completed with two twin 40mm, two twin and two single 20mm. Also fitted with Hedgehog on completion.

WILD GOOSE Completed with two twin and two single 20mm, later increased to four twin and two single 20mm. Hedgehog fitted on completion.

WOODCOCK Completed with two twin 40mm and two twin and two single 20mm. Hedgehog fitted on completion.

WOODPECKER Completed with two quadruple 2pdr and two twin and two single 20mm. Hedgehog fitted on completion.

WREN Completed with two twin and four single 20mm, increased to four twin and two single 20mm. Hedgehog fitted initially.

Radar

271 and 291 was standard fit for all ships; those intended for AA work, ie completed without Hedgehog, fitted 285 gunnery radar.

ACTAEON in full peace time guise. 291 at the masthead, 293 on a platform forward of the yardarm, 285 on the gun director above the bridge.

HMS ACTAEON U 07, F 07

Not completed until well after the end of hostilities, ACTAEON led a quiet and blameless life. Completed in 7.46, she followed the long established role for sloops by proceeding to a distant Station for detached duty, in her case the South Atlantic where she represented the Royal Navy in the West Indies and the South American coast generally, finally paying off to reserve at Portsmouth in mid 1953.

Transferred to Lisahally to lay up there in May 1956, ACTAEON was then one of the escorts purchased by the Bundesmarine as training ships in the first days of that Service's establishment and expansion. Sold to West Germany 11.11.57, and refitted in Britain, she commissioned on 9.12.58 as HIPPER.

Wright & Logan
ALACRITY, ultra smart and flying her paying off pennant, approaches Portsmouth after service in 3 FS. The 40mm amidships and aft are clearly visible, she also retains 20mm in the bridge wings. 293 at yardarm level has replaced 271, otherwise her wartime radar fit is unchanged.

HMS ALACRITY U 60, F 60

ALACRITY completed right at the end of the European war, so that Pacific service became inevitable, and she sailed accordingly from her post work up repairs on 19.6.45 for the Far East. Delayed by further exercises at Malta, and the inevitable minor defects, she did not reach Colombo until the war had come to an end, so that her first deployment was to proceed direct from Colombo to Hong Kong, where she arrived on 19.9.

ALACRITY operated off the Chinese coast in the ensuing months, representing the Royal Navy at Shanghai in late 9.45, and being based on Hong Kong. In the later part of the year, she joined the 32 EF at Hong Kong and served there until sent to New Zealand in 2.46 to undertake a refit in the dockyard at Auckland to mid 4.46. She then became part of the 1st Escort Flotilla, later becoming the Senior Officer's ship. On the Pacific Fleet becoming the Far East Fleet in 1949, ALACRITY remained in 1 EF until 5.49 when she transferred to the 3 Frigate Flotilla up to the end of 1951.

On arrival in Britain, ALACRITY reduced to reserve, at first at Portsmouth and then later at Lisahally from 1953, where she lay until approved for disposal. This was in 1956 when she was sold to BISCO and allocated to W.H. Arnott, Young & Co Ltd, arriving at Dalmuir 15.9.56.

The wartime AMETHYST in a censored view, examination shows that her pennants have been deleted. Twin 20mm in power mountings can be seen on the quarterdeck, with 40mm in the bandstands amidships. Her late model 271 is mounted at yard arm height, the only other radar is the 285 above the Director.

HMS AMETHYST U 16, F 116

AMETHYST on completion in 11.43 worked up at Tobermory and then, very briefly, joined 57 EG based at Greenock from where she made a round trip to the Mediterranean escorting troop convoys KMF 28 and MKF 28 to and from Gibraltar. Then transferred to 37 EG attached to the Mediterranean Fleet, AMETHYST commenced escort duty in that area which lasted until 10.44 when she returned home for a refit at Devonport, arriving 11.10.

On completion of refit, she proceeded north and joined 22 EG based on Liverpool, that Group operating in the Support role in the North Atlantic. Here AMETHYST drew blood when she took part in the sinking of U 482 on 16.1.45 and U 1276 on 20.2.45, continuing to cover convoys in the Western Approaches until the end of the European war.

A brief refit at Devonport 15.5 to 25.6 preceded the ship sailing for the Pacific, where she joined the Fleet at Manus in early 7.45. Operating in defence of the Fleet Train, AMETHYST then became involved in the immediate post war activities of the Fleet, and joined 32 EG in the first post war re-organisation.

In mid 1.46 she had a welcome break, being sent south to New Zealand to refit at Auckland until early 3.46, then to return to Hong Kong, the situation in China requiring a major part of the Fleet to be based there. During this period, AMETHYST was sent up the Yangtse to act as guardship and W/T link with the Embassy in Chungking, and it was while returning down river at the end of this duty that the well known incident involving the ship took place.

Fired on by the advancing Communist troops, who had by now occupied the north bank of the river, AMETHYST was trapped in the lee of a small island, with considerable casualties. Lieutenant-Commander Kerans, the Assistant Naval Attache, was sent down river to take command of the battered ship, the Royal Air Force in a very gallant action evacuated some casualties by flying boat while under heavy fire, and the balance of the ship's company settled down to an uncomfortable siege situation.

Realising that, should he remain, the ship would probably eventually be seized by the Communists, Lieutenant-Commander Kerans decided to break out from his dangerous berth and endeavour to gain the open sea and Hong Kong. In a night dash that caught the imagination of, at least, the British public, he succeeded in his endeavour and rejoined the Fleet at sea to make a triumphant return to Hong Kong.

While undoubtedly a gallant action, and justly applauded, one must wonder at the crass stupidity of Government and authority generally, that placed the ship in such a situation in the first place when the Chinese Communist reaction was almost inevitable and certainly predictable. Delusions of Imperial grandeur from days gone by can be the only answer.

WSPL

AMETHYST enters Plymouth Sound 1.11.49 to pay off, after her gallant dash down the Yangtse to evade the Communist blockade.

AMETHYST returned to a delirious welcome in Devonport, and paid off for refit, on completion of which she returned to the Far East at the end of 1950 as part of 3 FS for a further two years' service.

On return to Britain, she paid off at Devonport and was then laid up at Lisahally to await the inevitable end. However, after sale for breaking up, she achieved some notoriety as a "film prop" for she was hired from the breakers to play herself in a film of the Yangtse Incident. Filmed off Harwich, the special effects department achieved what the Chinese had failed to do - they holed the ship below the waterline with explosives during the filming and she commenced to fill. The efforts of a party of officers and seamen from the shore training establishment HMS GANGES were needed to save the ship from actually sinking, and the rather chastened film crew returned AMETHYST, somewhat battered and bent, to her owners to be taken to Plymouth, arriving there on 19.1.57 to be broken up.

WSPL

The end of AMETHYST. Stripped to main deck level, and with her bows already missing, AMETHYST lies in the breaker's yard at Plymouth.

A magnificent sunlit view of CHANTICLEER, which shows the eight DCT arrangement very clearly. She has twin 20mm in power mountings in the bandstands and single, manually operated, mountings in the bridge wings. 271 is on the stump tower aft, 291 at the masthead and 285 above the Director. The very clean state of the ship seems to indicate that the photograph was taken before commencing her Atlantic career.

HMS CHANTICLEER U 05

CHANTICLEER enjoyed what might be described as "a short life and a merry one" for, on completion, she joined 7 EG based at Greenock for duty in the North Atlantic generally. Her first escort could however be described as pedestrian, for she was one of several ships detailed to escort a large floating dock being towed south from the UK. This, with a speed of advance of some 5 knots, entailed the escort steaming in a constant circle round a temptingly slow target, and the eleven days spent until detaching to fuel at Lisbon must have been tedious in the extreme.

CHANTICLEER spent 6 to 9.43 escorting convoys to and from Gibraltar then, after brief defect rectification on the Clyde, entered the North Atlantic war in the support role, unfortunately without success. On 18.11.43, while she and her Group were hunting U 513, CHANTICLEER was unable to evade an acoustic torpedo and her stern was blown off. Still afloat, she was towed to Ponta Delgada by other ships of her Group; it says much for the efficiency of her complement that by the time she arrived all stores, ammunition, papers etc that could be removed were all packed and listed ready for distribution amongst the rest of the Group either for use or transport to Britain.

Paid off officially on 22.11.43, her presence was seized upon by the British Senior Naval Officer at nearby Horta who had a number of officers and ratings based ashore in less than desirable quarters, CHANTICLEER was accordingly towed to Horta by the salvage vessel *SALVEDA*. There the offices etc in the bridge structure replaced the destroyed officers accommodation aft, while the ship's services were restored so that power, heat and galley facilities etc became available for the base party who no doubt abandoned their quarters ashore above the Portugese Navy's main naval store with delight.

On 1.1.44 the hulk was re-named HESPERIDES and formally commissioned as a base ship at Horta, to serve as such for the remainder of the war. For some reason she was re-named again, as LUSITANIA, on 8.10.44 and so remained until paying off at the end of the European war, the hulk then being sold locally and finally broken up in Lisbon at some time after 1945.

A newly-built CRANE photographed in 5.43, with a slightly unusual complex camouflage pattern. There is the usual, standard, radar fit, there are power operated twin 20mm in the bandstands aft, and manually operated singles in the bridge wings.

HMS CRANE U 23, F 123

CRANE served from completion in 5.43 until 8.44 in 7 EG based on Greenock, almost entirely in the North Atlantic with much of that service in the support role. In 9.43 she was diverted briefly to A/S patrol in Biscay during the "blitz" of U boats making their passage across the Bay, but without success.

First blood was drawn on 18.11.43 when CRANE attacked U 515, her first success coming on 21.11.43 when she sank U 538. Further work in the North Atlantic followed until 8.4.44 when U 962 was sunk; shortly afterwards the Group was withdrawn to prepare for the blocking operations during the Normandy landings. These successfully concluded, CRANE was ordered to proceed from Devonport to the East coast via the Clyde for refit; unfortunately she collided with *TILAPA* en route on 7.8.44 and the resultant damage and the overdue refit lasted until 1.11.44, repairs being made on the Humber.

After repair, CRANE stored at Greenock and then worked up at Falmouth prior to sailing on 26.11 for the Far East as she was detailed to join the Pacific Fleet. En route she called at Brest, probably one of the first RN ships to do so since the end of the German occupation, and then went eastward via Malta, Aden,

CRANE at Spithead in 9.51. The split Hedgehog is clearly visible abaft B mounting, 293 has replaced 271 and is mounted at yardarm level with 285 and 291 retained in their original positions. Single 40mm are in the bridge wings and aft on the quarterdeck, with twin mountings in the bandstand.

Colombo and Fremantle. From Fremantle CRANE escorted the depot ships ARTIFEX and TYNE to Sydney NSW; arriving there 17.2.45. From Sydney she went on to the Fleet base at Manus, finally joining the Fleet there 4.3.45.

CRANE took part in the Sakashima Gunto Fleet Train operations in 5.45 and, at the end of hostilities, became part of 1 EG of the Pacific Fleet. Two months later she transferred to 31 EG and went to Brisbane for refit 17.10.45 until 1.46. CRANE remained with the Pacific Fleet, including a collision with the destroyer COCKADE 23.4.46, until 1.7.46 when she sailed from Hong Kong for home to pay off on arrival and lay up in reserve at Harwich.

In mid 1948 CRANE shifted her berth to Chatham, where she stayed until re-commissioned in late 1951 for service in 3 EF of the Far East Fleet where she remained for a number of years. Detached to the Red Sea for the Suez operations, CRANE was fortunate to escape serious damage when attacked by Israeli aircraft; although hit by rocket fire the impact fusing of the projectiles prevented penetration and consequent internal damage.

Returning to the Far East Fleet, CRANE continued her service in 3 EF with periodic refits and drafting of reliefs from the UK to Singapore, until she returned home for the last time in mid 1962 to pay off and lay up at Portsmouth, arriving at Queenborough to be broken up during 3.65.

For photographs of CYGNET see overleaf

HMS CYGNET U 38, F 38

CYGNET completed 1.12.42 and at once distinguished herself by grounding heavily on 9.12 during her work up. Repaired on the Clyde 19.12.42 to 14.2.43 she then spent a month at Tobermory for work up (much longer than usual) and briefly joined 2 EG at Liverpool for a return passage to Gibraltar with troop convoys.

In 4.43 CYGNET transferred to 7 EG based at Greenock and operated in the support role in the North Atlantic until 3.6.43 when she went into dockyard hands on the Clyde with defects, repairs that lasted until 24.6. Returning to escort duty, she then went to the wartime emergency dockyard at Dunstaffnage 24.9 for repairs that lasted until 26.12: it would seem that all was not well with this new ship. On completion of repair, a further Tobermory work up took place, immediately followed by a Russian convoy, JW 56A/RA 56 which must have tested the new ship's company.

CYGNET then returned to the North Atlantic, still with 7 EG, and had her first success on 8.4.44 when U 962 was sunk. Unfortunately, five weeks later, on 13.5 while entering Belfast, CYGNET grounded and spent until 13.7 under repair. Two more Russian convoys, JW59/RA 59A and JW 60/RA 60, followed with an Iceland ferry convoy as a change prior to a fortnight's attention to defects on the Clyde.

Russia then called again, for in 11.44 CYGNET commenced a series of three return passages to Russia which lasted until 1.3.45, ensuring a bitter winter for all onboard. Five weeks refit on the Clyde and at Corpach followed and then, Russia again! This time there was only one return passage with convoys JW 66/RA 66 during which U 307 was sunk on 29.4 prior to the sailing from Russia of RA66. The passage was followed by the inevitable defects period to make good weather damage etc, during which time the decision was taken to refit the ship at Leith from 1.6 to 3.9 for Pacific service.

CYGNET's ill luck held for, while on post refit trials she went aground and, by the time she completed repairs on 17.9 the war was over, and she went to Devonport to lay up, her crew no doubt looking forward to demobilisation.

In 1948 CYGNET joined the Fishery Protection Squadron, remaining with them until 1951 when she refitted for service with 5 FF in the Mediterranean, transferring to 2 FS in late 1952. In the late summer of 1954 she spent some time in the West Indies on loan to that Squadron and then returned to pay off and lay up at Chatham, where she remained until towed away 13.3.56 by the tug *MERCHANTMAN* to Rosyth, where she arrived 16.3.56 to be broken up.

HMS HART U 58, F 58

HART completed 12.43 and arrived at Tobermory in time to work up over Christmas, a pleasure the crew could no doubt have happily foregone. On completion, she joined the Clyde based 7 EG for support group duty in the North Atlantic, transferring to Gibraltar for similar duty in 3.44.

In 5.44, the Group returned to the Clyde to prepare for the blocking operations in the South West Approaches in support of Operation Neptune; during this build-up period HART collided with the sloop ROCHESTER, though damage appears to have been minor as she was on station on D day. Escort duty in the Channel area followed up to the end of 9.44 during which she collided with the troopship *VIKING* on 1.9.44.

Somewhat lengthy repairs were undertaken at Devonport 19.10-23.12, so that the ship arrived again at Tobermory for work up on Christmas Day 1944 (poor planning somewhere!) after which HART joined 22 EG based at Liverpool and soon shared the sinking of U 482 on 16.1.45. From her new base, escort for convoys in the approaches to the UK were the usual duty continuing until VE day, following which three weeks' repairs were carried out at Devonport to allow HART to go east to the Pacific Fleet, sailing from Britain 20.6.45.

CYGNET underway in the entrance to the Mersey, the picture gives great detail of the ship as built. The cramped quarterdeck with four DCT a side, 271 on its stump tower, two quadruple 2 pdrs in the bandstands. There are manual, single 20mm mountings in the bridge wings, but the starboard gun is not yet fitted.

An excellent impression of an escort in wartime guise.

A post war CYGNET, somewhat altered from the preceding views. Taken in the Portsmouth approaches, she is now Senior Officer, with a black funnel cap and no pennants shown. The 271 on its stump tower has gone, a pole mainmast now carries the 291 instead. Her DC chutes and stowages are empty, the sole close range AA armament is two twin 40mm in the bandstands. A lattice mast has replaced the tripod, and carries 293 in lieu of 271, while the elderly 285 has been retained.

A beautiful 1.44 picture of HART lying in the Clyde anchorage shortly after completion. Twin power mounting 20mm are evident on the quarterdeck and in the bandstands, manual operated singles in the bridge wings. 291, a late mark of 271 and 285 are the radar fit and Hedgehog can just be seen aft of B mounting.

Arrival in the Indian Ocean coincided with the Japanese surrender, so HART sailed from Colombo on 11.8 with orders to proceed via Singapore to Hong Kong to join the Fleet there. Having done so, a local re-organisation transferred her to 32 EF and sent her to refit at Brisbane 23.12.45-1.46 - for once the crew got Christmas Day alongside! Remaining with the Pacific Fleet, latterly in 1 EF, HART had a somewhat chequered career, grounding 14.4.46, ramming a jetty 13.3.47 and the freighter *HONG KHENG* 19.7.47, finally suffering typhoon damage in 9.49. The Fleet being re-titled Far East Fleet, HART transferred to 3 FF in 1949 and finally returned home to reserve at Devonport in mid 1951. At the end of 1953 the ship was towed to Lisahally to lay up there, returning to Devonport in 1955 and placed on the disposal list. Sold in 11.57 to the Bundesmarine and refitted in Britain, HART then became the FGS SCHEER on 24.1.58 for further service abroad.

Fotoflite incorporating Skyfotos

Except for the classic hull, this ship is probably unrecognisable but is in fact the German SCHEER ex HART. The addition of heavy block superstructure, a Germanic funnel cap and a number of complex radar aerials have completely changed her appearance.

A beautiful view of HIND fresh from the builders in 4.44. Power operated twin 20mm in the bandstands and on the quarterdeck, singles in the bridge wings. 285, 291 and 293 radars are also evident in this view.

HMS HIND U 39, F 39

HIND was one of the later sloops to complete during the war, finishing her Tobermory work up just in time to join the preparations for the Normandy landings, although not directly associated with any of the formed Groups. After D Day, she continued general escort duty in the Channel area until the end of 7.44, then went to Holyhead for the repair of defects prior to sailing to join the East Indies Fleet. HIND sailed eastward, escorting troop convoy KMF 35 as far as Port Said, and then collecting convoy AP 75 at Aden for passage to the entrance to the Persian Gulf.

WSPL

Whereas the previous photograph was the beginning, this is almost the end; a forlorn HIND lies neglected in Reserve awaiting disposal, the end for so many of the sloops.

Returning to Aden, HIND then took over another troop convoy, ABF 6 detaching from it with the Ceylon portion and arriving at Colombo 5.12.44. From there, she proceeded north to Calcutta and commenced duty escorting troop and store ships to and from Chittagong in support of the Burmese offensive.

With the re-organisation of the Fleet at the end of 1944, HIND joined 60 EG and remained on Indian Ocean duty until the end of 2.45 when unspecified defects kept her at Colombo. With the local repair facilities apparently unable to cope, HIND lay at Colombo until 7.4 and then limped back to Alexandria where she lay under repair from 23.4-1.8.45.

Seaworthy again, HIND was now allocated to the Pacific Fleet, although the war was over before she left Colombo on her way on 21.8.45. Joining that Fleet in the aftermath of a long war, she became part of 32 EF at the end of the year and refitted at Brisbane early in 1946. At the end of the year she refitted briefly at Singapore, and then moved to Hong Kong, returning to active duty at the end of 1947 in 1 EF and to serve in that flotilla until the re-naming of the Fleet as Far East Fleet and the Flotilla as 3 FF in 1949.

HIND returned home in 1951 to enter reserve at Portsmouth, moving to the Tees in 1954 to lay up at Hartlepool, where she was to remain until declared for disposal, arriving at Dunston on Tyne 10.12.58 to be broken up.

A magnificent study of KITE which, taken with the view overleaf, should satisfy any modelmaker requiring fitting details.

HMS KITE U 87

KITE completed in time to join the legendary 2 EG just at the turn of fortunes in the Atlantic battle, completing her work up at Tobermory 6.4.43. Her first operational sortie was as part of a support group sailing on 29.4, a style of work that lasted until mid 6.43 when she and her Group shifted to offensive patrolling in the Biscay area. This change of area brought her first success, when U 449 was found and sunk on 24.6.

A three week repair period at Devonport followed, and after sailing on 23.7 U 504 was found and sunk on 30.7. A further Biscay patrol followed, unfortunately without result, and KITE then returned to the central Atlantic.

On 13.10, while leaving harbour, KITE was in collision with the tug *ALLEGIANCE*; the minor damage was repaired at Londonderry in ten days. At sea again in support of convoys on 25.10, U 226 was sunk on 6.11, and KITE continued working the area until 17.12 when she commenced six weeks' repair at Liverpool to 28.1.44.

More support group work followed, which yielded U 238 as the victim on 9.2; this was to be her last success of the war. On completion of this cruise, KITE transferred to the Clyde based 7 EG and, on returning from her first cruise with that Group, collided on 30.3 with the freighter *ANNAGHMORE*, the damage being repaired on the Clyde from 1.4 to 11.5.

On sailing from the Clyde, KITE went south to Sheerness and thence to escort duty in the Channel in connection with the D Day traffic, work that occupied the ship until 3.8 when she went back to Scapa Flow to prepare for a Russian convoy. Sailing 15.7 with JW 59 KITE was unable, due to ASDIC conditions in northern waters, to detect the presence of U 344 on 21.8 and was hit amidships by two torpedoes. On such a small hull, the effect of two hits is both inevitable and rapid: KITE sank immediately and only nine survivors were picked up from the Arctic sea.

KITE very shortly after delivery in 3.43. This photograph was taken soon after the one on the preceding page.

HMS LAPWING U 62

LAPWING was another late building that entered service just in time to support the Normandy landing traffic, operating in and around the Channel from D Day until mid 9.44. Then being free to take up her more distant water role, she joined 7 EG based on the Clyde and started her career with that Group, which was to consist entirely of Russian convoy duty.

LAPWING was to make four round voyages to Russia with convoys between 20.10.44 and 28.2.45, heavy work even for such well-found ships. During this period she was in collision with *MOORBY* 28.11 when sailing with JW 62, and had the inevitable weather defects made good on the Clyde 10-26.11 and 24.1.-1.2.45.

LAPWING's last voyage began when she sailed on 11.3 with JW 65 for, when entering Kola Inlet after the convoy on 20.3, she was torpedoed, probably by U 968, and sunk.

One the book's oddities, and proof that even shipyard foremen can make mistakes. Obviously the foreman painter was told that LAPWING was "U pennants 62", and he had just that painted on the ship! Equally, being alongside in a basin, it is probable that the 1st Lt did not even see it prior to sailing, his comments (and those of FOIC Clyde) would have been interesting.

HMS LARK U 11

LARK completed and worked up in time to join the host of ships covering the Normandy landings, and remained on escort duty in the Channel area until 22.9, thereafter proceeding to the Clyde to make good defects. Operational once more, she then joined the Greenock based 8 EG and commenced a steady routine of Russian convoy escort, completing three round voyages prior to her loss.

Russian convoys took a heavy toll on all ships, including the most modern, and LARK repaired defects on the Clyde 10-25.11 after the JW 61/RA 61 passage, and 24.1-1.2.45 after JW 63/RA 63. She then sailed on her last passage on 3.2.45 and took JW 64 through to Kola arriving there 15.2. Sailing two days later, she attacked and sank U 425 on 17.2 but was herself torpedoed by U 968 (one of the most successful Arctic boats which also claimed LARK's sister LAPWING). Although not sunk, LARK was towed back in to Kola inlet severely damaged, and beached.

LARK, destined for a short life, displays her charms for the photographer shortly after leaving her builder's yard and going down to the Clyde Anchorage 4.44. It is interesting that by this time, 271 was ceasing to be fitted in sloops, 293 on a platform at yardarm level being the alternative.

The Senior British Naval Officer in North Russia had at his disposal the personnel of Naval Party 100 who were based at Murmansk and in the area to assist the convoy traffic. Having ascertained that it was unlikely that the ship could be made fit to be towed to Britain he commenced a rigorous stripping programme. By the time the RN workers had stripped the hull of all stores, spares and movable parts (including some items of main machinery) there was little left except a hulk, which was handed over to the Russians for disposal on 13.6.45.

Although there were persistent reports post war that LARK had been repaired and entered service with the USSR, even the name NEPTUN was reported, the massive stripping exercise carried out makes it very unlikely that this did in fact occur as the facilities available in North Russia in 1945-50 were quite inadequate to effect what would have amounted to a total rebuilding.

MAGPIE shortly after completion. She has the usual radar fit, 285, 271 and 291, Hedgehog abaft B mounting, and single 20mm in the bridge wings and bandstands.

Fotoflite incorporating Skyfotos
An immaculate MAGPIE comes up Spithead prior to the 1953 Coronation Review. Her radar suite is now the old 285 and 293; Hedgehog remains but the 20mm have been supplanted by single guns in the bridge wings and twin mountings abaft the funnel on each beam.

HMS MAGPIE U 82, F 82

MAGPIE worked up at Tobermory during 9.43 to join the Liverpool based 2 EG on completion and start work in the North Atlantic in the support role with that Group. It was not until 1.44, after a refit at Liverpool from 20.12.43-20.1.44 that the ship had any good fortune in her activities, but on 31.1.44 U 592 became the first victim, followed by U 238 on 9.2.

In 3.44 the Group was detached from Western Approaches to lend its expertise to Russian convoys, and sailed accordingly on 27.3 with JW 58, during which passage U 961 was sunk on 29.3. Returning with RA 58, MAGPIE repaired the inevitable weather damage at Liverpool 15 to 28.4 and then prepared for the Normandy landings. For this purpose the ship transferred from her Group and carried out convoy escort in the Channel itself after D Day, including the protection of the ships laying the PLUTO cable from Britain to France.

A brief refit at Chatham preceded her return to the North Atlantic, which she did by joining the newly formed 22 EG, hived off from the highly successful 2 EG a little earlier. Support operations for convoys in UK waters were now the order of the day as U boats moved ever closer inshore. Unfortunately MAGPIE had no further successes in the difficult hunting conditions that prevailed.

An eight weeks' refit at Portsmouth commenced 13.5 and the subsequent work ups were carried out in the more benign conditions of Gibraltar and Malta between 26.7 and 24.9 after which MAGPIE joined 33 EF of the Mediterranean Fleet, in which Fleet she was to serve until 1954.

In 1947 MAGPIE transferred briefly to 5 EF, then to 2 EF in which, under that title and 2 FF, she was to remain until late 1954. It was during this period that she became well known as the command of HRH Prince Philip, then still an active service Officer, whose wife lived ashore in Malta; one of the few times that Her Majesty has been able to enjoy something approaching what most people would regard as a "normal" life in pleasant conditions. MAGPIE returned home at the end of 1954 to join 6 FS in the Home Fleet, prior to just over a year's service on the South Atlantic Station. On completion of this, she paid off at the end of 1956 and was laid up at Devonport, being handed over to BISCO 9.7.59 and arriving at Blyth 12.7.59 to be broken up.

A nice post build view of MERMAID, listing heavily at anchor in the Clyde 4.44. Note the unusual two colour pennant, conforming to the camouflage pattern. Twin power mountings for 20mm fitted amidships and on the quarterdeck, manual singles in the bridge wings.

HMS MERMAID U 30, F 30

MERMAID was another "late starter", and appears to have suffered somewhat from teething troubles as her work up was a protracted affair commencing 2.6.44 and not completing, for various reasons, until 17.7. Even then the inevitable defects meant that she did not complete for service until 13.8 when she sailed with her first convoy, JW 59 to North Russia.

Perhaps as a result of all the training, she made her first sinking on 24.8 when the victim was U 354 and returning with RA 59A she sank U 394 on 2.9. MERMAID then made good defects and operated in and around the Clyde and Scapa Flow area until the end of 11.44. A lengthy refit at Leith then followed lasting until 23.3.45. After the inevitable Tobermory work up, MERMAID then joined 12 EG based on Liverpool. However, the European war ended very shortly afterwards, and she was allocated to join the Pacific Fleet. A refit at Portsmouth 14.5 to 17.7 was required to fit her for more tropical service than the Arctic and Scapa Flow, and the subsequent work up took place in the sunny climate of Malta for four weeks from 24.7.

MERMAID was then detailed for the slow job of escorting floating dock AFD 22 on passage from Malta to the Indian Ocean sailing on 28.8 for Aden. As the Japanese war was well over by the time Aden was reached, no further escort was required, neither was MERMAID needed in the Far East, so she returned to Malta and joined the Mediterranean Fleet, initially in 33 EF later 5 EF, then eventually in 1947, 2 EF.

Fotoflite incorporating Skyfotos
MERMAID, as Leader of 2 FF, steams to join the Coronation Review in 6.53, immaculate in new paint. A very clear photograph showing all her fitting details.

Her Mediterranean service was protracted; by 1951 she had transferred to 5 FF but within six months reverted to 2 FF which later changed its title to 2 FS. It was not until the latter half of 1954 that she finally came home to reduce to reserve at Portsmouth where she was to lie until disposed off by sale in 11.57 to the Bundesmarine, eventually to re-commission as FGS SCHARNHORST on 5.5.59 after refit.

The German SCHARNHORST ex MERMAID, in 7.59. Other than the cinder screen, beloved of the German Navy for many years, she is little altered since her sale.

After a further refit, SCHARNHORST is barely recognisable from her first post sale appearance, and certainly not as a British sloop design, only the basic hull remains unaltered in appearance.

courtesy M Louagie
After completion of her active career, SCHARNHORST was retained as a static Damage Control hulk for training purposes. She is seen here in 10.90 entering the canal leading to Bruges on her way to be broken up.

HMS MODESTE U 42, F 42

MODESTE was a late arrival, not completing until after the end of the Pacific war; in consequence her early career was limited, to say the least, by the constraints of finance and manning that affected the whole of the Fleet.

Following completion, MODESTE served briefly as the Gunnery Training ship and Firing ship attached to HMS EXCELLENT at Portsmouth, and then passed to reserve becoming part of the VERNON II

MODESTE on 12.9.45 in the brief interlude before she paid off to reserve at the end of the war. Note the single 40mm in bridge wings, amidships and on the quarterdeck, with two single 20mm mountings in the bandstands aft of the funnel. Unusually, she does not carry her pennants, possibly in view of her service as a tender to HMS EXCELLENT.

establishment, moored alongside the old battleship RAMILLIES and used as an accommodation ship for Sea Cadets. Even this employment ceased in 1950, and MODESTE remained idle until refitted in late 1952 for service in the Far East Fleet in 3 FS in 1.53, her first real service since completion.

MODESTE remained with the Fleet in the East until late 1958, with an interlude serving in the Red Sea during the Suez intervention, and then returned to Portsmouth to pay off and lay up, remaining in reserve until passing to BISCO 8.3.61, and arriving at St Davids on Forth 11.3.61 to be broken up after one of the shortest active lives of the class, losses excepted.

A somewhat spruced MODESTE sails from Aden 18.8.58 on her way home to pay off after service in 3 FS in the Far East. The carrier BULWARK and frigate BIGBURY BAY lie on the moorings beyond her.

A spruce NEREIDE 15.5.46, still very much as she was when built. Twin 40mm amidships, and singles in the bridge wings form the close range AA armament.

HMS NEREIDE U 64, F 64

Another post-war completion, NEREIDE was sufficiently late to be a brand new ship when the overseas squadrons were being reconstituted, so that she did not suffer the indignity of passing straight to reserve from the builders. Instead, late in 1946, she went to the South Atlantic Station where she was to remain, with interludes for refits and crew reliefs, until 1954. At the end of her time, she transferred briefly to the America and West Indies Station, and then came home to Portsmouth in mid 1954 to pay off and lay up. Unfortunately further service passed her by and she remained in the Portsmouth Reserve Group until scheduled for disposal, arriving at Bo'ness 18.5.58 to be broken up.

WSPL

A forlorn NEREIDE, already stripped of armament, lies in the breakers yard at Bo'ness in 1958 awaiting demolition.

OPOSSUM in the Clyde 6.45. She has a mix of close range weapons, single 20mm in the bridge wings, twin powered mountings aft, and twin 40mm amidships.

HMS OPOSSUM U 33, F 33

OPOSSUM completed in 6.45 and went East on completion of work up to join the escort force of the British Pacific Fleet, refitting at Sydney NSW in mid 1946. Following this came brief service in 1 EF of the Pacific Fleet, she then came home and laid up in reserve at Portsmouth in late 1947.

In 1951 OPOSSUM was scheduled for service in the Far East once again, with 3 FS which she joined, after refit, in the autumn of 1952 to serve in that Squadron until 1957. On leaving the Far East, OPOSSUM transferred to 7 FS on the South Atlantic Station and later on the South American Station, prior to going to reserve at Devonport during 1959. Her time in reserve was brief, for she arrived at the breaker's yard at Plymouth on 26.4.60 for final destruction.

WSPL

Not a particularly clear photograph, but shows OPOSSUM post 1947. 40mm singles on the bridge and quarterdeck, twins amidships; she is still very cluttered aft with a full DC outfit and equipment.

PEACOCK lying in a convoy anchorage 5.44, she is ready to sail, with full depth charge stowage on deck aft.

HMS PEACOCK U 96, F 96

PEACOCK had a typically abrupt introduction to escort duty, being sent to North Russia immediately after completion of work up on 21.7.44, and participating in the sinking of two U boats, U 354 and U 394, on the outward passage - on 24.8 and 2.9 respectively. Returning to Scapa Flow on completion of the return passage with RA 59 she spent a month based there and on the Clyde before going to Chatham for a six weeks' refit, completing 22.12.

PEACOCK then joined 22 EG based on Liverpool, the Group being employed in the support role in the Western Approaches; PEACOCK scored almost at once, taking part in the sinking of U 482 only six days after sailing on her first patrol. No further success came her way, despite active A/S work until the German surrender, and she then went to refit at Liverpool on 2.6, destined for service in the Pacific.

When PEACOCK completed her refit on 9.8 it was apparent that there was no need for further sloops in the Pacific, so her orders were amended to the Mediterranean Fleet, and she went to Malta to work up from 16.8 to 15.9. In the Mediterranean, in common with other escorts, a distasteful part of her duty was the interception of illegal Jewish immigrant carriers approaching Palestine; she stopped *DIMITRIOS* on 23.11.45 while on such a patrol.

1946 saw a re-organisation of the Fleet, and PEACOCK became part first of 33 EF then 5 EF and, finally, 2 EF. She was to remain with this Flotilla, re-titled 2 FF in 1949, until late in 1950 when she transferred to 5 FF but still with the Mediterranean Fleet.

1952 saw PEACOCK revert to 2 FS, with whom she saw out her time in the Mediterranean, coming home in mid 1954 to pay off and lay up at Portsmouth to await disposal. In fact, this did not occur until 1958 but on 7.5.58 she arrived at Rosyth to be broken up alongside the dockyard there.

Fotoflite incorporating Skyfotos
The post war PEACOCK steaming up Spithead 6.53 to join the Coronation Review, immaculate in a new paint scheme no doubt applied in the past few days in the lee of the Isle of Wight. Hedgehog can be seen abaft and to starboard of B mounting. Close range AA is limited to twin 40mm amidships.

PHEASANT in the Clyde anchorage 5.43 looking very new and smart. Single 20mm in the bridge wings and down aft, powered twin mountings amidships. She has 285, 271 and 291 radar.

PHEASANT was towed to Troon, arriving 13.1.63, to be broken up. However, she broke her tow en route and, prior to re-connection, it was deemed necessary to take off the RN working party onboard. Here a naval helicopter winches the men off while the dead ship lies stopped showing the "Not under control" signal.

HMS PHEASANT U 49, F 49

PHEASANT completed her work up, partly at Tobermory and later at Scapa Flow, on 12.6.43 and joined the Greenock based 7 EG then employed escorting convoys to and from the Mediterranean. During her first passage outward she grounded off Tripoli so that on return to the Clyde she spent from 2.8-27.9 under repair.

7 EG, including PHEASANT, now turned its attention to the North Atlantic and after one return passage to St Johns NF, operated in the support role until the end of 1943. PHEASANT herself went to refit at Newport, Mon. 1.12.43-24.1.44 and transferred to 55 EG of the Mediterranean Fleet, taking out convoy KMF 28A on her passage to the Station. Escort duty in the Mediterranean kept PHEASANT busy until 10.44 when she came home to refit at Leith from 12.10-7.12 prior to joining the East Indies Fleet.

PHEASANT escorted a convoy as far as Gibraltar on her outward passage and was delayed there by defects from Boxing Day to 13.1.45; during this time her orders were amended so that she was now to join the Pacific Fleet. Passage from Gibraltar to Manus via Colombo and Darwin was a lengthy affair: she finally joined the Fleet 5.3.45 and was allocated to the AA defence of the Fleet Train, with which she participated in the Sakashima Gunto operations.

After the end of hostilities, Fleet re-organisation made her part of 1 EG, later 32 EG, refitting at Brisbane from 19.10.45 to 2.46. Brief service in the Pacific followed and PHEASANT then returned to Portsmouth and paid off to reserve where she remained until 1954. By this time, the passage of time and lack of maintenance had lessened the utility of a recently constructed sloop so that she then had little value and was accordingly moved in late 1954 to the Barry group of the Reserve Fleet. PHEASANT lay at Barry, reverting to the control of the Plymouth Group in 1960 and was duly placed on the disposal list in 1962, handed over to BISCO 7.12.62 and arriving at Troon 15.1.63 to be broken up.

Almost a standard Clyde anchorage view of REDPOLE in 6.43 which shows these ships to their best advantage.

HMS REDPOLE U 69, F 69

REDPOLE led one of the more active lives of the wartime sloops, being almost continuously employed throughout her existence. Worked up at Tobermory and Scapa Flow, she joined the Greenock based 7 EG in 8.43 for a brief period before passing to the Mediterranean Fleet where she was principally Gibraltar based from 9.43-2.44.

Returning to Britain in 2.44, REDPOLE rejoined her former Group, 7 EG for a brief period which included covering the D Day landings and subsequent cross Channel traffic until 4.8.44. An eight week refit on the Clyde then preceded a transfer to the East Indies Fleet, REDPOLE escorting the troop convoy KMF 35 from Britain to Port Said on her passage east, finally arriving at Colombo for duty 6.11.44.

REDPOLE operated with the Fleet from Ceylon until 1.45, and was then ordered to join the Pacific Fleet, duly reporting for duty at Sydney NSW on 17.2.45 after a passage via Fremantle. Somewhat in need of refit by this time, she was then ordered to Auckland where she spent two months under repair prior to returning to Sydney, by which time the requirement for sloops on active service in the Pacific was lessening by the day.

Fotoflite incorporating Skyfotos

REDPOLE in her dis-armed state in Spithead, still a very good looking vessel. B and X mounting positions are now occupied by training bridges complete with compasses, chart tables etc for the student officers. 291 remains at the masthead, the lattice tower carries a 277 aerial, developed as the successor to the 271/3 series late in the war.

REDPOLE remained in the Pacific after the end of the war until 1946, when she came home and paid off to reserve, being laid up at Harwich until 1948. Transferred to Portsmouth during 1949, she was then stripped of armament and prepared for service as a Navigational Training Ship attached to the Navigation and Direction School, HMS DRYAD, near Portsmouth. REDPOLE served continuously in this role for ten years, operating out of Portsmouth and in UK waters training junior Officers in the arts of pilotage and navigation. Finally handed over to BISCO 16.11.60, she arrived at St Davids on Forth 20.11.60 for breaking up.

SNIPE in 9.46 ready for overseas service. She carries 293 in lieu of 271, and has twin 40mm amidships and singles in the bridge wings. As she is to serve on a small Station overseas, her pennants have not been painted up.

HMS SNIPE U 20, F 20

SNIPE was almost the last of the sloops to complete, leaving her builders 9.9.46 and ordered, after work up, to join the America and West Indies Station where she was to serve until 1952, with intervening refits and crew changes. As this Station included the whole of the American seaboard it was certainly a welcome change for her company from the austerity of post-war Britain.

In mid 1952 SNIPE was briefly attached to 6 FS of the Home Fleet and then, after the Coronation Review, went into reserve at Devonport and later Barry where she remained until arriving at Newport, Mon. 23.8.60 to be broken up.

SPARROW at the end of 1946. Like SNIPE, she does not carry her pennants, and possesses a similar close range armament. She has, presumably, just completed trials: a thorough painting seems overdue, marked as she is with the builder's yard dirt.

HMS SPARROW U 71, F 71

SPARROW was the last sloop to complete, in 12.46, and like her sister SNIPE entered service in 2.47 with the America and West Indies Station where she was to serve until 1952, with visits home for refit and crew changes. In late 1952 she re-commissioned for brief service in 3 FS of the Far East Fleet, but returned to the South Atlantic Station (which now covered Southern America) in 1954.

In mid 1956 SPARROW finally came home, and paid off for the last time, to enter reserve at Portsmouth where she lay until handed over to BISCO 22.5.58, finally arriving at Charlestown in tow of the tug *MASTERMAN* 26.5.58.

STARLING in 1943 sporting her Leader's black funnel top. 285 forward, a late model 271 aft and 291 at the masthead form the radar suite, twin 20mm amidships and singles in the bridge wings.

HMS STARLING U 66, F 66

STARLING is undoubtedly the most widely known sloop of the entire series, thanks not only to her brilliant wartime career under the command of Captain F J Walker, Royal Navy, but also to her long service as Navigation Training Ship at Portsmouth where she became known to many thousands of officers and men, and the public generally.

STARLING late war and post refit. She has lost the 271 and tower aft, stepping a light mainmast, 293 at yardarm level replaces the 271, 40mm amidships the twin 20mm, and she has now shipped a lattice foremast.

Completed in 4.43, STARLING was taken over by Captain Walker as the Senior Officer of 2 EG based at Liverpool. This appointment ensured that the ship would be at the heart of and totally involved in, all the activities of the premier A/S Group of the War. In addition to the fighting instinct of her Captain, and his undoubted expertise in the A/S field, the Group's service in the support role meant that any contact could be investigated to the end, there being no direct responsibility for a convoy to impose restrictions. In consequence, STARLING's sinking record of sixteen U boats is unsurpassed in the entire history of A/S warfare; although even she could not equal the achievement of the USS ENGLAND which succeeded in making all her six sinkings in a single day.

STARLING commenced her operational life 29.4.43 when she sailed from Liverpool, and her first victory did not come until 1.6 when U 202 was destroyed. 2 EG then shifted base to Devonport and joined the hunt for U boats on the Biscay passage routes where U 119 was rammed and sunk 24.6, putting STARLING into Devonport Dockyard for repairs until 3.8. On leaving Devonport, she was in collision with UMBRA, but this did not delay her cruise in the Bay of Biscay after which she and the Group returned to the Liverpool base.

Sailing again in the support role, STARLING then sank U 226 and U 842 on 6.11, this concluding her 1943 achievements; she went for a much needed repair period at Liverpool 17.12.43-27.1.44.

STARLING returned to the North Atlantic on 28.1; U 592 succumbed on 31.1, U 238 and U 734 on 9.2 and U 264 on 19.2, all in the course of one cruise. A further North Atlantic foray commencing on 7.3 yielded U 653 on 15.3, and STARLING and her Group then went to Scapa Flow, having been lent for duty with the Russian convoy JW 58. The difficult Arctic ASDIC conditions limited the Group's effectiveness, nevertheless U 961 was sunk 29.3, following which STARLING and her sisters returned to Liverpool to effect repairs.

Fotoflite incorporating Skyfotos
The ultimate STARLING, with training bridges and chart tables replacing her twin 4in, and classrooms built on aft, photographed in early 1957.

28.4 saw STARLING at sea again, and on 5.5 U 473 was the next victim; there was then a blank cruise from 29.5 to 1.7, which was to be the last conducted by Captain Walker. To the grief of his Group, Captain Walker died 9.7, and was followed in command of 2 EG by Commander Wemyss, who had been the next Senior Officer. In the first patrol under the "new management", the sinking of U 333 on 31.7, U 736 on 6.8 and U 385 on 11.8 showed that the old expertise continued.

STARLING went to refit at Falmouth 30.9-13.12.44 leaving 2 EG which was, in any case, divided to form a new Group, 22 EG which STARLING joined in 1.45. The touch had not left the ship however, for on 16.1 U 482 was sunk, her last success of the war.

On 12.2, STARLING went for a long refit on the Tyne with the intention that she would join the Pacific Fleet, and her work up at Tobermory 10.8 to 5.9 was so designed. The end of the war in the Pacific changed all that, and she went instead to Devonport to lay up, like so many other ships.

At the end of 1945, the decision was made to convert STARLING to a training ship, and she accordingly went into Portsmouth Dockyard where she was stripped and dis-armed for service with her sister REDPOLE as Navigation Training Ship, in which role she became a familiar sight in and around Portsmouth Harbour. Commencing her task 3.46, STARLING served continuously until 1959 when she passed to the Portsmouth Reserve Group to lay up.

STARLING was not placed on the Disposal List until 1963, and it was not until 6.7.65 that she arrived at Queenborough for disposal, after a long, active and highly successful life in both war and peace.

A sunlit WHIMBREL at anchor for a formal portrait. Twin 40mm amidships, single 20mm in the bridge wings, 285 radar forward and 271 aft, with HF/DF at the masthead.

HMS WHIMBREL U 29, F 29

WHIMBREL worked up at Tobermory in 2.43 and then joined the legendary 2 EG based on Liverpool under the command of Captain F J Walker RN. During 4 and 5.43 she operated with the Group in the North Atlantic completing one round voyage with convoys ON 170 and SC 123, then shifting to the Freetown route for a round passage with OS 46 and SL 129.

In 6.43, following three weeks' repair of defects on the Clyde, WHIMBREL joined 7 EG based on Greenock and commenced escorting convoys on the Gibraltar route until mid 10.42, with the variety of one Gibraltar to USA convoy, returning from America to Liverpool with a mixed troopship and fast tanker convoy, UT 1. Another round trip to Gibraltar was followed by a similar passage to Canada, after which WHIMBREL went to Hartlepool in early 11.43 for a refit lasting until the end of 1.44.

On completion of refit, WHIMBREL rejoined 2 EG at Liverpool 2.44 and thereafter operated in the support group role transferring from one convoy to another as the submarine threat varied. In 3.44 the Group, still operating as a specialised hunting formation, shifted its area of operations to the Arctic and was lent to the Home Fleet to escort a return convoy to North Russia, JW 58/RA 58. Not surprisingly, two weeks' repair were required at Liverpool on conclusion of that passage, which also served to prepare the ship for her part in the Normandy invasion.

In order to debar U boats from interfering in the Channel during and after the landings, WHIMBREL was one of many North Atlantic A/S specialists deployed in the entrance to the English Channel from early

June to the end of 8.44 with the sole object of denying the enemy access to the supply routes to Normandy, a duty that was admirably performed. On relief from the tedium of the intense patrolling, WHIMBREL repaired at Sheerness during 9.44 and, after a brief return to the Atlantic, was sent to Newport, Mon. to refit prior to proceeding eastwards.

Completing her refit 14.12.44, WHIMBREL was allocated to the British Pacific Fleet and worked up at home until 18.1.45 when she sailed for Malta and further exercises until 21.2. On completing training, the ship then went to Colombo, where she arrived 8.3, and onwards to Australia and Manus to join the BPF. In 5.45, with her powerful AA armament, she was one of the ships escorting the large Fleet Train during the Sakashima Gunto operations, and remained active in the AA force of the BPF until the end of the war.

WHIMBREL remained in the Pacific until the end of 1946 when she returned to Britain, paid off and lay in reserve at Harwich until the end of 1948. Transferring to the Medway, WHIMBREL remained there until 11.49 when she was sold to Egypt as part of a series of purchases to establish a small but relatively modern Royal Egyptian Navy, in which she served as a sloop under the name EL MALEK FAROUK.

courtesy Egyptian Navy
Almost unchanged, TARIQ in the 1980s. Russian close range weapons have replaced the 40mm but in similar positions, otherwise she retains her post war RN appearance.

Following the revolution that deposed King Farouk, the ship was re-named TARIQ (sometimes also transliterated as TARIK) in 1954 and remained in active service. Although, of recent years, she has spent most of her time alongside or at anchor as a training and accommodation ship, TARIQ still serves actively and is the last of the British sloop type still extant in almost the original form; indeed in 1992 plans were formulated to purchase the ship and repatriate her to Britain to join the growing fleet of preserved warships in this country.

WILD GOOSE in 5.46 configured for the Persian Gulf with extra cabins aft in lieu of X mounting. She mounts twin 40mm amidships only, and is white painted with a buff funnel, the Persian Gulf colours.

HMS WILD GOOSE U 45, F 45

WILD GOOSE completed in 3.43 and, after work up, joined 2 EG as the senior ship after the leader of the Group, STARLING. She served continuously in this Group throughout the war, succeeding as Senior Officer's ship on the death of Captain F J Walker RN, the Group Commander, in 7.44 and remaining under the command of Lt Cdr (later Captain) D E G Wemyss RN during that time.

By virtue of her continuous service in the most successful Escort Group of the war, WILD GOOSE had a lively and exciting time sinking, or participating in the sinking of, nine U boats. Between completion and transfer to Devonport in 6.43, she accounted for U 449 on 24.6.43, while operating in the Biscay area on offensive operations as opposed to convoy support, U 504 was accounted for on 30.7.43. Returning to the North Atlantic and convoy support U 842 was sunk on 6.11.43 which concluded her 1943 successes.

Repairs at Liverpool occupied from 18.12.43-27.1.44, allowing a Christmas alongside, and the sinking season then opened with U 592 on 31.1.44, followed by U 734 on 9.2.44, and by U 424 two days later. The next cruise yielded U 653 on 15.3 after which WILD GOOSE carried out a return convoy to North Russia, JW 58/RA 58, well away from her usual hunting grounds. No sinkings were made on that occasion, but on returning to the North Atlantic U 473 was put down on 5.5. A brief repair period at Liverpool in late 5.44 preceded the major A/S operations defending the Channel approaches during and after the Normandy invasion, and on 5.7 WILD GOOSE arrived at Belfast for an eight weeks refit.

The remainder of 1944 was quiet for WILD GOOSE, at least so far as successful attacks on U boats was concerned, despite the lengthy time spent at sea. Her final success in 1945 however, U 1208 on 27.2, concluded her wartime score.

The war in Europe ended, WILD GOOSE went to refit at Leith from 12.6 to 17.9.45 with the intention of reinforcing the Pacific Fleet, an intention nullified by the Japanese surrender. Instead, after a period laid up, WILD GOOSE proceeded to hotter climes, joining the Persian Gulf Division at the end of 1946 and served there for several years as the Senior Officer's ship, interspersed with absences for leave and refit.

Returning home to Chatham in early 1955, WILD GOOSE lay in reserve for less than a year, being sold in February 1956 and arriving at Bo'ness 27.2.56 for breaking up.

WOODCOCK in mid 43. Single 20mm in the bridge wings, twin 40mm amidships and what appear to be twin power mountings for 20mm aft. Still with a tripod, she mounts 285 forward and 271 on the lattice tower aft, 291 at the masthead.

HMS WOODCOCK U 90, F 90

Like several of her war-time sisters, WOODCOCK completed her work up and joined the 2 EG as one of a team of elite submarine hunters, although she had only one personal success, the sinking of U 226 on 6.11.43. Nonetheless, the ship formed part of a highly successful team which operated throughout the N Atlantic in support of convoys until the end of 1943.

On 4.12.43 WOODCOCK arrived at Dundee for repair; there she lay until 21.1.44 and then sailed to join 7 EG based on Belfast, also to operate in the Atlantic until 3.44 when she switched to the Gibraltar route at the beginning of the month. 4.44 saw her under repair at Belfast followed by a period based for local work on the Clyde prior to moving to Portsmouth in the build up to the Normandy landing.

On 27.5 she was, unfortunately, in collision with the destroyer VENUS leading to emergency repairs at Portsmouth, and on D Day itself she was just capable of proceeding from Portsmouth to Hull for permanent repairs that took until 7.8. Even these do not appear to have been totally satisfactory, for on 16.9 WOODCOCK arrived at Liverpool for further work which lasted until 8.12 including alterations to suit her for Pacific service. On 22.12 any ideas of Christmas at home, or indeed alongside, were shattered when the ship sailed for Gibraltar, spending Christmas Day in the Biscay area. A brief sojourn at Gibraltar was enlivened by a collision with the corvette BERGAMOT on 2.1.45, but this did not prevent WILD GOOSE from continuing her eastward journey so that she finally arrived at Manus to join the Pacific Fleet on 5.3.45 having travelled via Colombo and Darwin.

In the Pacific Fleet, modern AA armament and a modest speed decided that WOODCOCK, in common with the other sloops in the Fleet, should act as escorts for the Fleet Train and, as such, she operated in support of the Fleet during the Sakashima Gunto operations of 5.45.

Following the end of the Japanese war, WOODCOCK remained on station as part of the Pacific Fleet Escort Force until the latter part of 1946, when she returned home to lay up in reserve at Harwich, remaining there until the end of 1948 when she moved to Chatham in a high degree of readiness for service. This was gradually reduced as time passed, and by mid 1953 the ship had moved to Hartlepool, where there was a large laid up group. WOODCOCK never left this, finally being disposed of, arriving at Rosyth 28.11.55 in tow of the tug *SUPERMAN* to be broken up.

WOODPECKER coming up the Mersey with Perch Rock Battery in the background as she approaches the Gladstone Dock. She has single 20mm in the bridge wings, twin 20mm aft and quadruple 2 pdrs amidships. Radar suite is 285, 271 and 291.

HMS WOODPECKER U 08

WOODPECKER was another of those unfortunate ships whose career was brief and exciting. She sailed from her Clydeside building yard on 18 December 1942 for work up at Tobermory and then service with the legendary 2nd Escort Group; service that was to last little more than one year.

Completion of work up, and a period alongside at Londonderry for defects and storing, preceded her first trans Atlantic convoys with ON 164 and SC 120, then WOODPECKER moved to the Gibraltar route with troop convoys KMF 11 & MKF 11; service that lasted until early 4.43. The emphasis then shifted to the support role, and WOODPECKER so operated until 9.6, unfortunately without result.

Shifting to the Biscay area, WOODPECKER and 2 EG came under Devonport Command and joined the "Bay offensive", otherwise titled Operation Musketry, which involved close RN/RAF co-operation against submarines transiting the Biscay area. WOODPECKER's first success came on 24.6 with the sinking of U 449 prior to returning to Devonport to fuel etc.

A further blank patrol followed, but on the third such cruise U 504 was sunk on 30.7. WOODPECKER then required dockyard attention, and spent almost three months in hand at Avonmouth.

Returning to 2 EG and Western Approaches Command on completion, escort of KMF 26 and MKF 26 to and from Gibraltar during December preceded a return to support role patrols commencing in 1.44.

Return to the support as opposed to escort function brought both reward and disaster; U 762 was sunk on 8.2.44, U 424 on 14.2.44 and U 264 on 19.2. The following night U 256 struck with an acoustic torpedo which neatly removed WOODPECKER's stern, depriving her of all motive power and steering. Remarkably, there were no casualties onboard neither was the submarine sunk by the Group

STARLING succeeded in taking her stricken sister in tow at dawn and summoned the tug STORMKING to assist, handing over the tow on her arrival; WOODPECKER's crew having spent the time shoring up and generally preparing the ship to the best of their ability.

WOODPECKER would have reached Devonport had it not been for the weather; nearing the Scillies on 26.2 the weather worsened and just after 0800 on 27.2 WOODPECKER capsized and sank. Her survivors were picked up by the corvettes AZALEA and Canadian CHILLIWACK and brought into Devonport.

WREN 10.5.49 in the Mediterranean. She is now reduced to two single 40mm in the bridge wings, and four 3 pdr saluting guns amidships, apart from the main armament. Extra cabins have been added aft, with a light mainmast in lieu of the 271 tower.

HMS WREN U 28, F 28

WREN completed her work up at Tobermory and joined 2 EG 5.3.43 thereafter carrying out convoy escort on the Gibraltar route and on the North Atlantic run until 6.43. The Group as a whole then deployed to Devonport to undertake offensive A/S operations in the Biscay area, Operation Musketry, during which WREN sank U 449 on 24.6. Her second patrol drew a blank, but during the third she found and sank U 504 on 30.7; the two subsequent voyages were again devoid of success.

Returning to their Liverpool base 21.9, WREN was struck by defects and obliged to refit at Avonmouth from 15.10 to 18.12.43, followed by a further Tobermory work up. On completion, she rejoined 2 EG and undertook a series of support group passages in the North Atlantic until 24.3.44. Arriving in Scapa Flow on that date, the Group, including WREN, escorted the North Russian convoy JW 58/RA 58 before returning to their North Atlantic hunting grounds, where WREN found and sank U 473 on 5.5.

Used during 6.44 to prevent access by U boats to the D Day supply lanes, WREN sailed again on a support patrol 8.7 and sank U 608 9.8 shortly before arriving back at Liverpool. There was then a brief repair period at Falmouth during 9.44 and a further patrol in the North Atlantic before a longer refit on the Humber from 20.10.44-27.1.45.

On completion of refit, WREN joined the 22 EG and spent the remainder of the European war in patrols and support activities in the Western Approaches, unfortunately without success. WREN then had a one month refit at Cardiff during 4.45, and a longer time at Rosyth from 9.6.45 preparing for Eastern service. The end of the war cancelled these plans and, after completion, WREN in fact proceeded to the Persian Gulf at the end of 1946.

Service in the Gulf, interspersed with refits at Colombo and Malta, continued until mid 1955 when WREN returned to the UK and paid off to reserve at Portsmouth. Here she lay until approved for disposal, being towed from Portsmouth by the tug *MERCHANTMAN* and arriving at Rosyth 2.2.56 for breaking up.

RIN Modified BLACK SWAN class

Name	Builder	Laid down	Launched	Completed
CAUVERY	Yarrow	28.10.42	15.6.43	21.10.43
KISTNA	Yarrow	14.7.42	22.4.43	26.8.43

Displacement 1,925 tons standard. Dimensions length 229ft 6in, beam 38ft, draught 11ft 4in at full load.
Machinery two shaft, geared turbines, designed SHP 4,300 = 19.75 kts. Oil fuel 390 tons, consumption 0.8 tons per hour at 10 kts.
Initial armament three twin 4in HA, four twin and two single 20mm, Hedgehog.
Radar.
271 and 291 fitted on completion.

Changes during service
CAUVERY Four single 40mm post war, also four 3 pdr saluting guns.
KISTNA Two single 40mm, two twin and two single 20mm, also four 3pdr saluting guns.

CAUVERY, probably lying at Tail o' the Bank. She has twin power mounting 20mm in the bandstands and on the quarterdeck, single manual mountings in the bridge wings. The Director carries 285, while 271 is on a stump tower aft; the Adcock aerial of HF/DF is at the masthead.

HMIS CAUVERY U 10, F 110

CAUVERY completed her work up at Scapa Flow and then escorted an outward and inward convoy before going to Tobermory for a further work up session. On completion of this she was ordered to join the Eastern Fleet, and duly sailed 3.3.43 escorting troop convoy KMS 29A as far as Port Said where she arrived 16.3. From Port Said CAUVERY collected a further convoy, AJ 2, at Aden which was escorted to Colombo, arriving 4.4.

Based on Colombo, CAUVERY operated on the eastern seaboard of India until 7.44, when she became part of an A/S hunting group, Force 66, searching the western Indian Ocean between Ceylon and Kilindini for long range U boats. On completion of these operations, without result so far as she was concerned, CAUVERY went for refit to Bombay in mid 11.44 lasting until mid 2.45.

Returning to active service, CAUVERY operated off Akyab, took part in Operation Dracula (the occupation of Rangoon) and patrolled extensively in the Andamans area. She was one of the early arrivals at Singapore after the surrender and went as far east as Manus towards the end of the year.

On the partition of India and Pakistan in 1947, CAUVERY remained in Indian service, eventually being employed as Cadet Training Ship and her name being amended to the more correct rendering of KAVARI, as the English equivalent of the Indian script. She was eventually paid off 30.9.77 and presumably broken up in India.

CAUVERY, or KAVERI as the name became rendered in Indian Navy service, seen late in her career. 40mm are now mounted in the bandstands, and a deckhouse has replaced X mounting. 271 and 285 radars have been removed, and replaced by what appears to be an RN 293. HF/DF has been retained.

HMIS KISTNA U 46, F 46

KISTNA completed her post work up defects at Londonderry 28.10.43, and then operated from Greenock escorting convoys to and from Gibraltar until the end of the year. Ordered to Indian waters, she acted as part escort to convoy KMS 36 to Port Said, and then convoy AB 30 from Aden to Bombay, arriving there for the first time 15.2.44. There followed a short period on the Bombay/Persian Gulf/Aden route until the end of 3.44, KISTNA then took convoys from Bombay to Colombo and on to Calcutta to commence escort duty on the Army supply line from Calcutta to Chittagong. This duty lasted from 12.4 to 29.5, and there was then a short period of service on the Calcutta to Colombo route, followed by transfer to the Bombay base.

India's KISTNA under way at speed in her original form and with full A/S armament.

A lovely view of KISTNA in 1961 when she had been reduced to a training ship, with accommodation aft in lieu of X mounting. As a Cadet Training Ship she made many overseas visits, here she is in East African waters with her crew and trainees at Harbour Stations.

By the end of the year KISTNA had returned to Ceylon and in 1.45 moved to the Burmese coast where the RIN became increasingly involved in supporting the Army, now moving south. In mid 3.45 she went back to Bombay for a six week refit and then returned to the Trincomalee base to begin patrols around the Andaman Islands in 6 and 7.45. After the Japanese surrender, KISTNA called at Penang and then returned to the main RIN base at Bombay.

Post war, KISTNA remained in Indian service after 1947, as a training ship and for auxiliary duties when finally past operational deployment. The transliteration of her Indian name was, latterly, rendered as KRISHNA. She finally paid off at the end of 1981, to be broken up by the Indian scrap industry.

List of abbreviated titles of ship builders and engine makers

Where a ship has been engined by a company other than the builder, the name of the engine manufacturer appears after the oblique stroke following the builder's name.

Abbreviated title	Full title
Chatham	Royal Naval Dockyard, Chatham
C Laird	Cammell Laird & Co Ltd
Cockatoo	Cockatoo Island Shipbuilding & Engineering Co
Denny	William Denny & Bros
Devonport	Royal Naval Dockyard, Devonport
Fairfield	Fairfield Shipbuilding & Engineering Co Ltd
Furness	Furness Shipbuilding Co Ltd
H Leslie	R & W Hawthorn Leslie & Co Ltd
J Brown	John Brown & Co Ltd
R Westgarth	Richardsons Westgarth & Co Ltd
S Hunter	Swan Hunter & Wigham Richardson Ltd
Scotts	Scott's Shipbuilding & Engineering Co Ltd
Stephen	Alexander Stephen & Sons Ltd
Thornycroft	J I Thornycroft & Co Ltd
White	J Samuel White & Co Ltd
Yarrow	Yarrow & Co

The Devonport Dockyard - built FALMOUTH at Bombay in 1943.

PENNANT NUMBERS

As with all warships, sloops were allocated a pennant number on completion consisting of a Flag Superior and two numeral pennants. Unlike most other small warships, these were not painted on the hull, probably due to there being so few sloops on any given Station. From 1926 to the major change early in 1940, the Flag Superior was **L**; after April 1940 this changed to **U** which was peculiar to sloops. From 9.39 until 1947, with few exceptions, pennants were painted on the hull abeam the bridge and across the counter. In 1947, with the general re-organisation and reclassification of ships, all sloops used **F**, ships retaining their original numbers EXCEPT: CAUVERY became 110, AMETHYST became 116 and CRANE became 123. The pennant numbers appear after each ship's name in the individual histories; the list below is therefore in numeric order for use in the "decode" form.

01	BRIDGEWATER		50	ROCHESTER
03	ERNE		51	MILFORD
05	CHANTICLEER		52	GODAVARI
07	BITTERN (L), ACTAEON (U)		53	DEPTFORD
08	WOODPECKER		56	ENCHANTRESS
10	CAUVERY		57	BLACK SWAN
11	LARK		58	HART
12	SANDWICH		59	LOWESTOFT
15	FOWEY		60	ALACRITY
16	GRIMSBY, AMETHYST		61	AUCKLAND
18	FLAMINGO		62	LAPWING
20	SNIPE		64	NEREIDE
21	JUMNA		65	WELLINGTON
22	FOLKESTONE		66	STARLING
23	CRANE		67	INDUS
25	SCARBOROUGH		69	REDPOLE
27	HASTINGS		71	SPARROW
28	PENZANCE, WREN		72	WESTON
29	WHIMBREL		73	WARREGO
30	MERMAID		74	SWAN
32	SHOREHAM		75	EGRET
33	OPOSSUM		76	LONDONDERRY
34	FALMOUTH		77	YARRA
36	LEITH		80	HINDUSTAN
38	CYGNET		81	STORK
39	HIND		82	MAGPIE
40	NARBADA		84	DUNDEE
42	MODESTE		86	PELICAN
43	BIDEFORD		87	KITE
44	PARRAMATTA		90	WOODCOCK
45	WILD GOOSE		95	SUTLEJ
46	KISTNA		96	PEACOCK
47	FLEETWOOD		97	ABERDEEN
49	PHEASANT		99	IBIS

SLOOP DISPOSALS

In view of the relatively small number of ships concerned, ultimate fates are given below in a common alphabetical table, additional to the data contained in individual histories.

ABERDEEN	arrived Hayle 19.1.49 to bu by T W Ward Ltd.
ACTAEON	sold W Germany 11.11.57, commissioned 9.12.58 as HIPPER, hulked 31.7.64 and sold 25.10.67 to Eisen & Metall KG, Hamburg for bu.
ALACRITY	arrived Dalmuir 15.9.56 to bu by W H Arnott Young & Co Ltd, hulk arrived Troon 3.11.56 for final demolition.
AMETHYST	arrived Plymouth 19.1.57 to bu by Demmelweek & Redding, after previous sale for use by film company in the Orwell.
AUCKLAND	lost to air attack off Tobruk 24.6.41.
BIDEFORD	to BISCO 14.7.49 and bu at Milford Haven by T W Ward Ltd.
BITTERN	damaged by air attack off Namsos, and sunk by JANUS 30.4.40 in deep water.
BLACK SWAN	arrived Troon 13.9.56 to bu by West of Scotland Shipbreaking Co Ltd.
BRIDGEWATER	used for bomb trials, to BISCO 22.5.47 and broken up at Gelleswick Bay by Howells.
CAUVERY	renamed KAVARI in Indian service, paid off 30.9.1977 and sold for bu 1979.
CHANTICLEER	torpedoed by U 515 18.11.43, constructive total loss and became a depot ship at Horta, being renamed HESPERIDES, later LUSITANIA; broken up at Lisbon in 1946.
CRANE	arrived Queenborough during 3.65 to bu by Lacmots Ltd.
CYGNET	arrived Rosyth 16.3.56 in tow of tug MERCHANTMAN to bu by Shipbreaking Industries Ltd.
DEPTFORD	to BISCO 8.3.48, arrived Milford Haven 11.5.48 to bu by T W Ward Ltd.
DUNDEE	torpedoed by U 48 14.9.40 in convoy SC 3.
EGRET	sunk by glider bomb in Biscay 27.8.43.
ENCHANTRESS	sold 22.10.46 for £22,500 as LADY ENCHANTRESS for commercial service, subsequently arrived Dunston on Tyne 16.2.52 in tow of the tug ENGLISHMAN to bu by Clayton & Davie Ltd.
ERNE	after service as RNR drillship WESSEX at Southampton, sold and towed away to Antwerp 27.10.65 having been sold to General Navigation and Storage Co of that port.
FALMOUTH	after service as RNR drillship CALLIOPE at Newcastle, arrived Blyth 30.4.68 to bu by Hughes Bolckow Shipbreaking Co Ltd. Sold for £11,480.
FLAMINGO	sold to W Germany and commissioned as GRAF SPEE 21.1.59, struck from active list 1964 and sold 25.10.67 to Eisen & Metall KG, for bu.
FLEETWOOD	arrived Gateshead on Tyne 10.10.59 to bu by C W Dorkin & Co.
FOLKESTONE	after bomb trials, to BISCO 22.5.47 and bu at Gelleswick Bay by Howells.
FOWEY	sold in 10.46 for £9,500 to Wheelock Marden & Co Ltd for commercial service as FOWLOCK. Broken up at Mombasa during 1950.
GODAVARI	became Royal Pakistan Navy SIND in 1947, sold 2.6.59 for bu in Pakistan.
GRIMSBY	lost by air attack 25.5.41 off Tobruk.
HART	sold W Germany and commissioned as SCHEER 24.1.58. Paid off in 1967 and sold 17.3.71 for bu to VEBEG, Hamburg.

HASTINGS	to BISCO 2.4.46 and bu at Troon by West of Scotland Shipbreaking Co Ltd, arriving for bu 10.4.46.
HIND	arrived Dunston on Tyne 10.12.58 to bu by Clayton & Davie Ltd.
HINDUSTAN	to Royal Pakistan Navy in 1947 as KARSAZ, bu in Pakistan during 1951.
IBIS	lost by air attack 10.11.42 during Operation Torch.
INDUS	lost by air attack 6.4.42 off Akyab.
JUMNA	renamed JAMUNA and paid off finally 31.12.80 being bu in India.
KISTNA	renamed KRISHNA and paid off finally 31.12.81 being bu in India.
KITE	torpedoed and sunk by U 344 21.8.44 with Russian convoy JW 59.
LAPWING	torpedoed and sunk probably by U 968 20.3.45 with Russian convoy JW 65.
LARK	torpedoed by U 968 17.2.45 off Kola with convoy RA 64, towed in, beached and stripped; formally paid off 13.6.45, hulk being handed over to the Russians for disposal.
LEITH	sold for commercial service 25.11.46 as *BYRON* later re-named *FRIENDSHIP*. Purchased by Royal Danish Navy in 1949 as GALATHEA, bu in 1955.
LONDONDERRY	arrived Llanelly 8.6.48 to bu by Rees Shipbreaking Co Ltd.
LOWESTOFT	sold for commercial service 4.10.46 as *MIRAFLORES*, bu in Belgium at Boom, near Antwerp, arriving 5.8.55.
MAGPIE	arrived Blyth 12.7.59 to bu by Hughes Bolckow Shipbreaking Co Ltd.
MERMAID	sold W Germany and commissioned as SCHARNHORST 5.5.59. Hulked as D C School at Neustadt 1974, arrived Zeebrugge for bu 24.4.90.
MILFORD	to BISCO 3.6.49, arriving Hayle 27.7.49 to bu by T W Ward Ltd.
MODESTE	arrived St Davids on Forth 11.3.61 to bu by J A White & Co Ltd.
NARBADA	to Royal Pakistan Navy as JHELUM 1947, sold for bu in Pakistan 15.7.59.
NEREIDE	arrived Bo'ness 18.5.58 to bu by P & W MacLellan Ltd.
OPOSSUM	arrived Plymouth 26.4.60 to bu by Demmelweek & Redding.
PARRAMATTA	torpedoed 27.11.41 and sunk by U 559 off Tobruk.
PEACOCK	arrived Rosyth 7.5.58 to bu by Shipbreaking Industries Ltd.
PELICAN	arrived Preston 29.11.58 to bu by T W Ward Ltd.
PENZANCE	torpedoed by U 37 24.8.40 with SC 1.
PHEASANT	arrived Troon 15.1.63 to bu by West of Scotland Shipbreaking Co Ltd.
REDPOLE	arrived St Davids on Forth 20.11.60 to bu by J A White & Co Ltd.
ROCHESTER	arrived Dunston on Tyne 14.2.51 to bu by Clayton & Davie Ltd.
SANDWICH	sold at Bizerta 8.1.46 for £3,050.
SCARBOROUGH	arrived Thornaby on Tees 3.7.49 in tow of tug *AIRMAN* to bu by Stockton Shipping & Salvage Co.
SHOREHAM	sold for commercial service as *JORGE F EL JOVEN* 4.11.46, broken up 11.50 at Bruges.
SNIPE	arrived Newport, Mon. 23.8.60 to bu by J Cashmore Ltd.

SPARROW	arrived Charlestown 26.5.58 to bu by Shipbreaking Industries Ltd.
STARLING	arrived Queenborough 6.7.65 to bu by Lacmots Ltd.
STORK	arrived Troon 3.6.58 to bu by West of Scotland Shipbreaking Co. Ltd.
SUTLEJ	finally paid off 31.12.78 and broken up in India
SWAN	finally paid off 21.9.64 and broken up at Sydney NSW by Hurley & Dewhurst during 1965-66.
WARREGO	finally paid off 8.8.63 and broken up at Sydney NSW by Hurley & Dewhurst during 1965-66.
WELLINGTON	sold 6.2.47 and converted to use as a Livery Hall on the Thames, still extant.
WESTON	after bomb trials, to BISCO 22.5.47 and bu at Gelleswick Bay by Howells.
WHIMBREL	sold to Egypt 11.49 for further service as EL MALEK FAROUK, renamed TARIQ and still in service in 1993.
WILD GOOSE	arrived Bo'ness 27.2.56 to bu by P & W MacLellan Ltd.
WOODCOCK	arrived Rosyth 28.11.55 to bu by Shipbreaking Industries Ltd.
WOODPECKER	torpedoed by U 256 20.2.44, capsized and sank 27.2.44 while in tow in Western Approaches.
WREN	arrived Rosyth 2.2.56 in tow of tug *MERCHANTMAN* to bu by Shipbreaking Industries Ltd.
YARRA	sunk by Japanese cruisers ATAGO and TAKAO while covering a Javanese evacuation convoy 4.3.42.

SUMMARY

Lost by submarine attack	6
Damaged by submarine and unrepaired	2
Lost to air attack	6
Lost to surface attack	1
Sold for further naval service	5
Sold for commercial service	5
Sold for use as a Livery Hall	1
Sold for scrapping	44
Extant in 1993	1
Total	71

SINKINGS OF SUBMARINES

The following sinkings are credited to sloops, either totally or as a participant ship, and are listed in date order.

IMPORTANT

Certain of the sinkings here listed differ in detail from previously published records, and some losses previously credited to ships are now omitted. These amendments are based on the latest research carried out on original documents, British and foreign, by the Naval Historical Branch, Ministry of Defence (Navy), and are up to date to 31 December 1992.

GERMAN SUBMARINE LOSSES

Date	U-boat	Ships
30.1.40	U 55	FOWEY, WHITSHED and a/c
31.5.40	U 13	WESTON
5.4.41	U 76	SCARBOROUGH, WOLVERINE
19.10.41	U 204	ROCHESTER, MALLOW, CARNATION
17.12.41	U 131	STORK, EXMOOR, BLANKNEY, STANLEY, PENTSTEMON and a/c
19.12.41	U 574	STORK
21.12.41	U 567	DEPTFORD
6.2.42	U 82	ROCHESTER, TAMARISK
14.4.42	U 252	STORK, VETCH
11.7.42	U 136	PELICAN, SPEY, Free French LEOPARD
31.7.42	U 213	ERNE, ROCHESTER, SANDWICH
2.4.43	U 124	BLACK SWAN, STONECROP
6.5.43	U 438	PELICAN
11.5.43	U 528	FLEETWOOD and a/c
1.6.43	U 202	STARLING
14.6.43	U 334	PELICAN, JED
24.6.43	U 119	STARLING
24.6.43	U 449	WOODPECKER, WILD GOOSE, WREN, KITE
15.7.43	U 135	ROCHESTER, BALSAM, MIGNONETTE
30.7.43	U 504	WOODPECKER, WILD GOOSE, WREN, KITE
30.8.43	U 634	STORK, STONECROP
1.11.43	U 340	FLEETWOOD, ACTIVE, WITHERINGTON and a/c
6.11.43	U 226	STARLING, KITE, WOODCOCK
6.11.43	U 842	STARLING, WILD GOOSE
21.11.43	U 538	CRANE, FOLEY
31.1.44	U 592	STARLING, MAGPIE, WILD GOOSE
8.2.44	U 762	WOODPECKER

Date	U-boat	Ships
* 9.2.44	U 734	STARLING, WILD GOOSE
* 9.2.44	U 238	STARLING, KITE, MAGPIE
*11.2.44	U 424	WILD GOOSE, WOODPECKER
19.2.44	U 264	STARLING, WOODPECKER
15.3.44	U 653	STARLING, WILD GOOSE and a/c
29.3.44	U 961	STARLING
8.4.44	U 962	CYGNET, CRANE
14.4.44	U 448	PELICAN, SWANSEA
5.5.44	U 473	STARLING, WILD GOOSE, WREN
31.7.44	U 333	STARLING, LOCH KILLIN
6.8.44	U 736	STARLING, LOCH KILLIN
9.8.44	U 608	WREN and a/c
11.8.44	U 385	STARLING and a/c
12.8.44	U 198	GODAVARI, FINDHORN
24.8.44	U 354	PEACOCK, MERMAID, KEPPEL, LOCH DUNVEGAN
2.9.44	U 394	MERMAID, PEACOCK, KEPPEL, WHITEHALL and a/c
16.1.45	U 482	HART, PEACOCK, STARLING, AMETHYST, LOCH CRAGGIE
17.2.45	U 425	LARK, ALNWICK CASTLE
20.2.45	U 1276	AMETHYST
27.2.45	U 1208	WILD GOOSE, LABUAN, LOCH FADA and a/c
29.4.45	U 307	CYGNET, LOCH INSH, COTTON

*The sinkings of U 734, U 238 and U 424 are confirmed. What is subject to further investigation is which submarine was sunk in which attack.

ITALIAN SUBMARINE LOSSES

Date	Submarine	Ships
23.6.40	EVANGELISTA TORRICELLI	SHOREHAM, KINGSTON, KANDAHAR
23.6.40	GALVANI	FALMOUTH
13.12.42	CORALLO	ENCHANTRESS

JAPANESE SUBMARINE LOSSES

Date	Submarine	Ships
11.2.44	Ro 110	JUMNA, LAUNCESTON, IPSWICH

INDEX

To ships which form the subject of this book. Principal entries only are indexed, i.e. where a ship is mentioned incidentally in another ship's history this mention is not indexed.

ABERDEEN 6, 7, 21, 42, 43, 118, 119
ACTAEON 7, 20, 82, 83, 84, 118, 119
ALACRITY 7, 82, 83, 84, 118, 119
AMETHYST 7, 19, 82, 83, 85, 118, 119, 123
AUCKLAND 6, 7, 16, 67, 68, 118, 119
BELVOIR 10
BIDEFORD 6, 7, 32, 33, 118, 119
BITTERN 6, 7, 15, 21, 61, 62, 63, 118, 119
BLACK SWAN 6, 7, 17, 76, 77, 118, 119, 122
BRIDGEWATER 6, 7, 11, 23, 118, 119
BYRON 51, 120
CALLIOPE 39, 119
CAUVERY 7, 114, 118, 119
CHANTICLEER 6, 7, 82, 83, 87, 118, 119
CRANE 6, 7, 19, 82, 83, 88, 118, 119, 122, 123
CYGNET 2, 6, 7, 82, 83, 89, 118, 119, 123
DEPTFORD 6, 7, 42, 45, 118, 119, 122
DUNDEE 6, 7, 38, 118, 119
EGRET 6, 7, 67, 68, 118, 119
EL MALEK FAROUK 110, 121
ENCHANTRESS 6, 7, 15, 21, 61, 62, 63, 118, 119, 123
ERNE 6, 7, 76, 78, 118, 119, 122
FALMOUTH 6, 7, 38, 39, 118, 119, 123
FLAMINGO 6, 7, 18, 76, 79, 118, 119
FLEETWOOD 6, 7, 14, 21, 42, 46, 118, 119, 122
FOLKESTONE 6, 7, 26, 118, 119
FOWEY 6, 7, 12, 32, 34, 118, 119, 122
FOWLOCK 34, 119
FRIENDSHIP 51, 120
GALATHEA 51, 120
GENTIAN 9
GODAVARI 6, 7, 15, 71, 118, 123
GRAF SPEE 80, 119
GRIMSBY 6, 7, 42, 47, 118, 119
HART 7, 20, 82, 83, 89, 118, 119, 123
HASTINGS 6, 7, 26, 28, 118, 120
HERON 7
HESPERIDES 87, 119
HIND 7, 82, 83, 92, 118, 120
HINDUSTAN 6, 7, 12, 31, 118
HIPPER 20, 84, 119
IBIS 6, 7, 76, 81, 118, 120
INDUS 6, 7, 14, 60, 118, 120
JAMUNA 73, 120
JHELUM 75, 120
JORGE F. EL JOVEN 37, 120
JUMNA 6, 7, 15, 21, 71, 72, 118, 120, 123
KARSAZ 31, 120
KAVARI 114, 119
KISTNA 7, 19, 114, 115, 118, 120
KITE 6, 7, 82, 83, 93, 118, 120, 122, 123
KRISHNA 116, 120
LADY ENCHANTRESS 64, 119

LAPWING 6, 7, 82, 83, 94, 118, 120
LARK 6, 7, 95, 119, 120, 123
LEITH 6, 7, 42, 49, 118, 120
LONDONDERRY 6, 7, 42, 51, 118, 120
LOWESTOFT 6, 7, 14, 42, 52, 118, 120
LUSITANIA 87, 119
MAGPIE 6, 7, 82, 83, 96, 118, 120, 122, 123
MERMAID 7, 20, 82, 83, 97, 118, 120, 123
MILFORD 6, 7, 13, 38, 39, 118, 120
MIRAFLORES 53, 120
MODESTE 7, 82, 83, 99, 118, 120
NARBADA 6, 7, 15, 16, 71, 74, 118, 120
NEREIDE 7, 82, 83, 101, 118, 120
NONSUCH 7
NYMPHE 7
OPOSSUM 7, 82, 83, 102, 118, 120
PARRAMATTA 7, 55, 56, 118, 120
PARTRIDGE 7
PEACOCK 6, 7, 82, 83, 103, 118, 120, 123
PELICAN 6, 7, 16, 67, 69, 118, 120, 122, 123
PENZANCE 6, 7, 26, 29, 118, 120
PHEASANT 6, 7, 19, 82, 83, 104, 118, 120
REDPOLE 6, 7, 82, 83, 105, 118, 120
ROCHESTER 6, 7, 32, 34, 118, 120, 122
SANDWICH 6, 7, 11, 23, 24, 118, 120, 122
SCARBOROUGH 6, 7, 12, 26, 29, 118, 120, 122
SCHARNHORST 20, 98, 120
SCHEER 20, 91, 119
SHOREHAM 6, 7, 32, 36, 37, 118, 120, 123
SIND 72, 119
SNIPE 7, 19, 82, 83, 106, 118, 120
SPARROW 7, 19, 82, 83, 107, 118, 121
STARLING 6, 7, 18, 82, 83, 107, 119, 121, 122, 123
STORK 6, 7, 15, 61, 64, 118, 121, 122
SUTLEJ 6, 7, 15, 71, 75, 118, 121
SWAN 7, 55, 56, 118, 121
TARIQ 110, 121
WARREGO 7, 55, 57, 118, 121
WATERHEN 7
WELLINGTON 6, 7, 13, 16, 19, 42, 53, 118, 121
WESSEX 79, 119
WESTON 6, 7, 38, 41, 118, 121, 122
WESTON-SUPER-MARE 7, 41
WHIMBREL 6, 7, 16, 82, 83, 109, 118, 121
WILD GOOSE 6, 7, 16, 82, 83, 110, 118, 121, 122, 123
WOODCOCK 6, 7, 16, 82, 83, 111, 118, 121
WOODPECKER 6, 7, 16, 82, 83, 112, 118, 121, 122, 123
WREN 6, 7, 16, 22, 82, 83, 113, 118, 121, 122, 123
WRYNECK 7
YARRA 7, 55, 59, 118, 121